The Topological Imagination

•

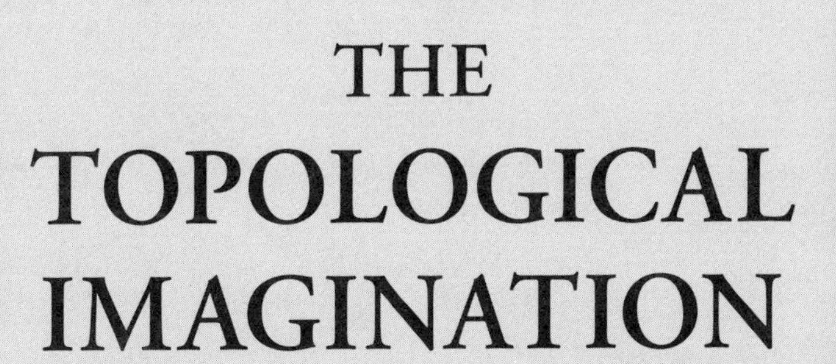

THE TOPOLOGICAL IMAGINATION

Spheres, Edges, and Islands

ANGUS FLETCHER

Harvard University Press

Cambridge, Massachusetts, and London, England

2016

Printed in the United States of America

First printing

Library of Congress Cataloging-in-Publication Data

Fletcher, Angus, 1930– author.
The topological imagination: spheres, edges, and islands / Angus Fletcher.
pages cm
Includes bibliographical references and index.
ISBN 978-0-674-50456-1
1. Topology. 2. Mathematics—Philosophy. 3. Art and science. 4. Knowledge, Theory of. I. Title.
QA611.F495 2016
514—dc23 2015032348

To Michelle

Contents

Introduction

If we humans are obsessed by the global, it is not without cause—our successful commerce joins with interlocking, essentially tribal wars whose bitter results are more than mind can tolerate— and yet at this moment in history much of our technology implies greater and greater cooperation among rival interests. The United States manufactures and exports weapons, but we also improve and send agricultural produce and other aid all over the world, contradictory actions prompting me to wonder about the right choice of expository method, if in some measure I am to address the global system in coherent global terms. Perhaps for me an informal style is appropriate, and it is not so strange to be writing a personal, ethically focused essay about an impersonal machinery of hyperactive business, war, and power. The task then becomes to describe one sort of unfamiliar consciousness, a seriously neglected kind of thinking, if the world today seems too large for comprehension or too small for analysis, and yet that foray into the unknown charts my conjectural voyage.

It is not clear how, reasonably, if we are to diminish global anxiety, we might stabilize the objects of our thought, or putting it another way, how should we humans estimate the *quality* versus the *quantity* of human consciousness? Semantic issues play a major role here. Perhaps we err in thinking that the words of ordinary language and the numbers employed in scientific measurement are equivalent to an approximate degree, as if the word "three" were radically identical to the number "3," and convenience alone gives rise to a strictly mathematical language. These may seem odd and unnecessary questions about language in the broadest sense, but they relate deeply to our human subjectivity. I, for example, am fascinated by the horizon afforded by a geometric sphere, and with such interests in mind I have gathered my conjectures around one center, the topological imagination, a gathering grounded in a special mathematical procedure: topology. In one sense this volume has been a shared meditation on our human home, the sphere itself, in all its dreams and divisions. In another sense it has simply been a wandering personal story about looking, listening and reading as a way of life shaped by the arts of two different languages.

To begin with the mathematics, we know that our human species lives on the surface of a large spherical surface, an irregular one and yet a manifold not totally unlike the perfect sphere a geometer would recognize. The ideal geometrical sphere has two sides (inside and outside) and no edges. Our planet has several interconnected layers of insulation, such as the atmosphere and stratosphere, lithosphere and hydrosphere, while many of its natural surface features create *de facto* edges, giving to our home in space an extremely complex global form. We are not supposed to have any edges, but we see edges everywhere, and islands of virtually spherical isolation are scattered everywhere on the planet, and magically they, too, seem to violate ideal uniformity.

My essay explores different ways of seeing the terrestrial biosphere, a *living space* revealed by what is called *topology,* where the placement of our lives is an imagined quality of our existence. That sounds remarkably abstract, when speaking of custom and human life, but there is no other way, for the mathematics enforces the problem of mixing and matching some patently realistic situations with some highly abstracted conceptions, unavoidably if we are to obey a virtually dramatic probability. We are like the two lovers in a famous lyric by the seventeenth-century poet Andrew Marvell, who imagined the origins of all yearning—"Had we but world enough and time, / This coyness, lady, were no crime." We dream, for we do not know how to participate in all four dimensions at once. The spherical itself, as the mind intuits its almost tactile quality, presents to us an impression of continuous change, and because the determining shape is that of a round ball, we live, usually without quite knowing it, in a state of continuous flux. It is said that Heraclitus the ancient Greek philosopher observed, "everything flows," and yet there is still some kind of order within the flow, an order coming precisely from the constant presence, everywhere, of differentiating edges between different things and different events and scenes. If our sphere could talk, it would say, "I am always looking for useful edges and trying to get rid of useless ones. Good edges, clear like longitude and latitude, will show me how to navigate on an otherwise boundless surface."

In a Heraclitan sense topology is a way of seeing and understanding the forms of Marvell's "world enough and time," as if they were constantly in motion, as if the mind could actually touch their motion, making bodily and hence physical contact with its abstract contours, by recognizing the soul of what is materially real, no matter how much the outward appearances of things may alter over time.

Hidden in the grand rhetoric of what a modern critic might call the dramatic monologue of the poem, "To his Coy Mistress" confronts a series of puzzles: how indeed could we measure the magnitude of world and time, and could humans ever know what is "enough" of either? How is the turbulence of life and its understanding to be kept smooth enough to be tolerable? At what ethical price can the world avoid braking its tumult into radical fragmentation? How, finally, can life have a flowing continuity? These are ancient yet no less modern questions, in their paradoxical mode that my essay pursues throughout, for we have made an accelerated world threatened by temporal *discontinuities*. Topology, we shall see, speaks directly to this enigmatic continuous phasing of long-term human survival. More than ever, without becoming regressive troglodytes, we must achieve order during the rise of radical and widespread *Mutability,* as four hundred years ago the Elizabethan poet, Edmund Spenser, called it, when the West had already entered the dawn of modern science and modern politics. Our question once again will have to be: how shall we think of permanence in the midst of fundamental change? The West went through a frightening hundred years of crisis, the seventeenth century, before it could begin to answer that question in abstract terms. It was only a beginning, however, and the Enlightenment period was to see revolutionary developments.

My starting point will therefore be two major mathematical discoveries made in the eighteenth century by a remarkable mathematician, Leonhard Euler, and I shall be returning many times to his thought, without claiming to analyze later development in mathematical fields. As a scholar working with literature and the arts my competence in mathematics is, of course, limited, and I am chiefly interested in the elective affinity, to extend Goethe's phrase, through broad analogies linking topology with artistic endeavor. Furthermore, the two inaugural eighteenth-century moments have implications for thought *in general,* again apparent in the links between art and science.

In 1735 Euler solved a picturesque local puzzle involving the famed Seven Bridges of Königsberg, and with that solution he established *analysis situs,* which we call topology. The story of this discovery did not end here, however. No less significant was Euler's second topological insight: His *Polyhedron Theorem* of 1750 showed how objects of a certain kind are endowed with a lasting and stable form, owing to an invariant relationship which establishes basic continuities, no matter how variable or different particular such objects may appear to the naked eye. On this ground, for example, an unsliced Idaho potato can be "the same mathematical object" as a perfect cube of dry ice, or a billiard ball or fisherman's harpoon.

Euler's twin discoveries depended on a radical insight. He stressed the site or placement of such objects, their positioning in space, including the way they actively locate their various internal parts, producing a unity we may call a complete form. To give a more recent example of such positioning, modern geochemistry examines the surface manifolds and the internal shaping of biochemical topology and thereby envisions a principle of organic form and biotic facts corresponding to Erwin Schrödinger's famous lecture, "What Is Life?" This essential question recently reappeared in a book-length expansion upon Schrödinger, identically titled *What Is Life?,* by Lynn Margulis and Dorion Sagan. Among other issues, the authors include a massive shift of analytic scale downwards to the microbial level where, surprisingly, Euler's new mathematics had everything to do with our human vitality, since biological shape as place is virtually defined by proportions between parts and powers working together as a kind of thermodynamic machine. The smallest living creatures participate in the life of much larger animals, let's say, the whale or the hippo, and all such life-forms belong to a vast combinatorial union called the environment. It follows that to understand life we must be concerned with the way its local character coheres within a wider terrestrial location, as the Earth travels in the cosmos, by moving up and down varying scales of power and value. Eventually that concern allows us to think about just and equitable scales of life's resources on planet Earth—not elsewhere, but here on the planetary surface of our sphere.

Here too we find the need for abstraction (perhaps an unavoidable impediment at the beginning of a book). One of the most difficult things about thinking through topology is that it treats space as radically abstract, as pure and even disembodied outline; the science involves "mathematical objects" rather than their obvious physical correlates—its sphere is a purely relational abstraction, whereas a soap bubble or a soccer ball are real things with physical attributes. We have even more trouble imagining that space itself, appearing to be inherently empty, can have a shape, for how can one bend or stretch a void of nothingness? It defies common sense to picture space-time being warped and curved, and yet it is so, because our universe is a cosmic dance of masses and energies all together conspiring to mold this emptiness we call space, where the abstract choreography eludes us, for we are accustomed to the experience of physical objects such as tables and chairs, rays of light coming from a lamp, water tumbling down a mountain stream, or the Moon, which in a sense give us the wrong idea about space. Strangely, we deprive the spatial emptiness of its true energy when we fail to consider its transformations, which are almost, we may say, the signs of life.

Coming down to earth, change in state occurs everywhere and under myriad conditions, as when we say that the attitudes or even the character of a person may change, let's say when he or she gets a promotion, though the person remains the same individual, recognizable in most cases, and I may give examples from the arts, to suggest how stretching the "literal" truth enters our lives in a topological fashion. The arts explore the serious meanings of this stretching of common literal reference and so it seemed, once again invoking the spirit of Andrew Marvell, that the enclosure of a seventeenth-century garden became a metaphysical experience,

> Annihilating all that's made,
> To a green thought in a green shade.

Such is the power of elegant form, wedded to the power of an idea. The arts demonstrate that if we seek to understand the boundaries and borders or even the clouded shaping of things in the world, we shall have to combine the science of topology with the methods and attitudes of art, both disciplines requiring intense imaginative agility. Despite ubiquitous change, everywhere and always, art and the topologist perceive an underlying stability, and they take the measure of this invariance. Both art and science show us that ideal and theoretical concepts may fall short of the crude truth of life, where, let us say, no sphere is a perfect mathematical object, but only an approximation.

Euler dramatically perceived an identity between place and thing, things being a kind of place or location of the real. In this light, place may display a changing shape, from pyramid to sphere, for example, yet still retain a topologically invariant stability of form. We then find ourselves at once asking who decides the borders of places, and who decides ownership—what is this or that thinglike place, in any case? To what, as much as to whom, does it belong? In what sense is a position an object? What kind of body am I seeing or thinking about? Such is the interest of this conjectural essay.

No concern of this kind, which the Ancient Greeks expressed in the grammar of the *middle voice* (verbs used for actions in which we take an active *interest*), could be more critical to actions taken across the present global stage. When political borders are redrawn on the map, *despite evidence that the land is essentially the same,* topologically no change of shape has occurred, yet we can hardly avoid the obscurity here: is a change in quantity (a redrawn border, assigning territory to new owners) not necessarily a change in quality? An armed border is not the same as an unarmed border, and the nature of the bounded body reminds us of an ancient historical riddle, whose basis I shall later discuss in relation to *The New Science,* Giambattista Vico's epochal eighteenth-century treatise—*plus ça change, plus c'est la même chose.* The proposition is not so much that Earth or other similar large bodies, or the human body for that matter, are always changing but not changing, as that our idea of metamorphosis itself will be seen to have changed, as if we suddenly realized that ideas are the shapes of mind. My essay therefore treats primarily how we imagine change in process, as perceived flow, and only secondarily suggests problems of measuring the transformations. In the midst of global fracturing often enhanced in its quarrels by the newly established digital technology of communication, an instantaneous flow, there is a crying need to understand what is meant by connectedness, by the novelist E. M. Forster's phrase, "Only connect." This task is by no means an academic exercise, for massive global changes now call insistently for more control than policy permits, or at least a finer general sense of the ethical need to reshape our unexamined thoughts.

In a world where, in addition, exponentially accumulating knowledge conflicts with the need for plain understanding, let us then posit a cluster of parallels drawn from different disciplines, which correspond to major themes in the book: *initial principles of topology; a broad-based poetic theory of the imagination; an account of the sphere without edges; a cyclical theory of history; ecological aspects of edge and division in the global context; a brief account of certain edge-types; an approach to ethics considered as a matter of scale, and finally an approach to islands and the emergent situation.* Integrating these fields has not been easy: while each chapter or section of the book focuses on questions of imagination from a particular perspective, the overall scope of intersecting approaches implies a single interpretive interest, whose aim is to reach a reasonable, coherent account as it might have been casually interwoven by Michel de Montaigne in one of his late essays on human behavior. For him the term *essayer* means to experiment with conceptual errors and trials; decisively he rejects all notions of final decision, and his exploratory style implies that discovery must at times proceed in the dark, feeling one's way, like an abandoned miner touching the jagged wall from one point to the next to the next. The glimmer of light is always at hazard, and there is never enough time.

The reader should associate my reasonings on the plane of analogy, given that all parts of the essay combine with each other through metaphors and converging analogical connections, which widen but also may intensify a shared communal sense of appropriate relevance, as if we were to say, "My *interests* may seem unlike yours, but we share biotic intentions that move across the material boundaries separating us. Together, despite our knowing the global magnitudes of size, numbers, directions, and distances (we use the satellite technology of GPS, sometimes to avoid thinking), we are still learning to fit our mental landscape to shifting perspectives, where solid terrain may yet guide a sense of style, in that way honoring the shared living biosphere, however degraded by humans it may be."

Technicians sometimes professionally question this picture of purposes as utopian, for they are specialists driving advanced engineering on many fronts and have reason to be uncomfortable with wandering thoughts whose precision, whose cash value—in William James's phrase—is hard to fix, although we sense its importance for understanding value itself. Wonderment as to why we humans are so predatory remains the cultural problem, however, precisely because consequences are so often unintended, and probably we need to think not only *as if,* living in the now, which means acquiring a feel for imprecision. We need approximating trends, one might say, instancing the fact that the globe is an object of long-range study at this time, as exemplified by neutral scientific analysis of global warming. The specialist's problem in taking the measure of things—let us say, for example, the trends of atmospheric changes and their causes—is located in this very difficulty, navigating the obscure passage in knowing how to cross the borders of theory and measurement with safe return, as if every improvement calls for an exit strategy.

Showing its colors most clearly in mathematics, precision itself always takes many forms, like the real world itself. Consider simple walking or the golf swing of an expert golfer, which differ radically from precise settings of the Large Hadron Collider in Switzerland. From organic to inorganic we cross a vague area, perhaps we could say from shape to measurement, and yet there is exactitude in either case. When it comes to human behavior and human conditions, as in quantum mechanics, our precision must be statistical and in that way approximately measuring things with sufficient accuracy. Dealing with contradictions on all sides, we start by using models, proceeding gradually to more exact description, but even then great puzzles may remain, as with the physicist's enigmatic double slit wave / particle experiment.

Topology in this regard is not unusual, when obeying the demand for accuracy, since the basic topological notion of *transforming a shape* is in fact only possible to describe by means of controlled analogy, as I see it. If shape is in some way alien to measure, even in music, I accept the analogical approach throughout this essay, in full knowledge that a mathematician would want a differently articulated sort of book. So also would an expert in reading Shakespeare wish for something beyond a supposedly correct gloss, and I shall not hesitate to cross both specialist boundaries, in the hope of sharpening and deepening a general sense of meaning for the intelligent reader.

The topological attitude is what matters for my account of imagination. To begin analogically, a topic familiar to me from many years of literary study would be the establishment of a "good text." What is the true shape of *King Lear*?—does it exist in quite dissimilar alternate versions? In various perspectives one would like to know exactly which text of *Hamlet* is closest to the words the poet may have written for its production, or indeed for different productions at the Globe Theatre. Details in searching for the better text are sometimes wildly complicated. Similarly, scientists have shown, especially since the development of quantum mechanics, that Nature seems to refuse an honest, obvious, authoritative physical text, as if we were always confronted by what the Ancients called a *sorites,* the "paradox of the heap," as R. M. Sainsbury calls it in his wide-ranging book on logical paradoxes. Every next grain of sand in the heap of variables introduces the need for a clearer theory of counting the uncountable.

In this transitional period we humans are all citizens of an empire of quantity, where there is too much to be counted and calculated, at least so it appears. Yet the heap is just the beginning, for clear values to be ascribed to meanings—having mentioned the play, let's say we are rather like its hero, Hamlet, who told his mother, "Seems, Madam? I know not seems." The State of Denmark was a poem in the hero's mind and in his author's teeming language. Perhaps everything we say or make or count is similarly a momentary seeming in the vast emptiness of time. And yet we count and yet we continue to measure the quantity of things to be dealt with. However precisely we attempt to gloss the words and meaning of nature (Galileo called the figures of geometry its "language"), and no matter how vigorously our observations and mathematics carry technology forward, these advances leave us humans exactly what Shakespeare's aging monarch, King Lear, had said we are; in our final, unsophisticated, impoverished state we are "bare, forked animals." We continue to live within starkest natural limits, and they evoke the Why secreted below the surface of the How.

In this context we may prefer a limiting goal, a refinement leaving us with the opposite goal of casting our nets widely, but not too widely. As I say in different ways throughout this essay, there is a need for perspective transformations, as an experimental psychologist like James J. Gibson might put it, analyzing *perceptual edges;* otherwise we shall not grasp the problem of fair discriminations between roughly similar things. Nor need we go back to the ancient myth of Daphne and Apollo, where, to escape her seducer, she prays for complete bodily change, and her arms and hands are turned into the branches and leaves of the laurel tree. In the Baroque period Bernini's wondrously carved marble statue, those arms and fingers are no less, perhaps more real than ever were their "originals." Art thus brings the viewer into immediate contact with riddles of the real and their relation to points of origin.

To clarify the large question of bounded transformations—the great Roman poet Ovid called them *Metamorphoses*—I emphasize edges, boundaries, borders, contours, and disparities, many of them measurable, others elusive and radically requiring the art of metaphor, a mental and verbal and pictorial skill related to what we may call the landscape of thought. We must build our theories to suit the terrain of the objective world, since our pursuit of relevant horizons belongs to the field of the *architectonic,* to use an old word from the text of Sir Philip Sidney's *Defense of Poetry,* defending the Renaissance. Metaphorically, there is an architecture or design in all things, and we encounter this early in life. One of our first learning experiences is to discover that shape involves a kind of self-evidence; to perceive a shape, of a table for example, is to see something as if its identity (though not its use) were self-evident. This morphology we experience on many scales of being, including, when we go to school, the wide range of recorded human memory.

Such orders of historical change, when following their shifting yet always human continuity, are the topic of Section Six in this essay, where I consider the historicist work of the Neapolitan scholar Giambattista Vico. He is important for two main reasons, first because he argued for a cyclical theory of history, but also because he believed that we humans can only understand and know what we have made. This principle (called *verum factum*) includes our making of theories as well as making things, our technical skills as well as our social customs, all of which exist in the mode of poetry, or as the ancient Greeks would have said, a *poesis.* Like a reborn Herodotus, Vico gave back to Western thought an invigorated understanding of intersecting historical and cultural cycles, which are essential for ecological knowledge and imaginative poetics—the source, Vico believed, of civilization in general.

Analyzing the hybrids of complex research has become an urgent priority in our time, and to illustrate this, the historian of science Professor Norton Wise compiled studies from different fields in 2004. His collection, *Growing Interpretations,* includes a detailed commentary by Peter Galison on the competing stresses between the physics and the mathematics of string theory; Evelyn Fox Keller writes on the debated concept of "self-organization"; David Aubin chronicles the conflicting attitudes and the tumultuous post-1950s history of René Thom's topological catastrophe theory, while other leading authors write about subjects as different as immunology and computer-driven artificial life. Endless new developments are occurring in adjacent and remote fields of science and cultural studies, including economics, where throughout we perceive a search for general coherence. Norton Wise is concerned with what he calls "boundary crossings" between many complex disciplines. For him, as many techniques develop, their discoverers inhabit a dynamic period of uncertain flux, where energy and matter are seen by scientists to interact on shifting levels of higher and lower structural stability (2004). Such, I would argue, are the stresses and bridges and edges and crossings on which my own essay is designed to shed light, especially in their relation to patterns of a much larger historical interpretation of culture, since our thoughts are always embedded in history, so far as we faintly understand historical time.

I

Topology and the Idea of Form

As a mathematical discipline topology makes a deep connection with idealist philosophy and ultimately illuminates ethical thought, and at the same time its methods animate the most exacting research into fields such as neurophysiology and computer science. To clarify this point of departure let us insist that topology and geometry, although connected by virtue of their abstract application to space and spatial relations, are radically different in spirit, as their names indicate. Among the ancient Greeks the term *geometry* literally meant "measurement of the earth," the prefix *geo* having the same root as the name for the earth-goddess, *Gaia,* whereas topology, *analysis of site* or geometry of position, as it was called for a considerable period, means "the logic of place." Henri Poincaré spoke always of *analysis situs,* when penning his five great essays on algebraic topology, which he virtually invented toward the end of the nineteenth century. For him the Latin *situs* and the much older Greek *topos* were identical, whereas Euclid's *Elements* are measuring *sizes, angles, areas, and spaces as measures,* while topology avoids measurement of sizes and similar magnitudes.

We notice, however, frequent disregard for etymology when we find that early topologists commonly used the word *geometry* to refer to place and *topos.* The refusal is not completely curious, though it might seem to violate the mathematical principle of precision, if one wishes instead to focus on shapes, even vague shapes which do not lend themselves to measurement, at least in nature, where knowing the diameter of a redwood tree cannot possibly convey its shape to the eye. Nonetheless, as we shall see, there is nothing vague about topology in its analyses, where the focus on place turns out to be a focus on shapes, which are in fact *dis-positions.* Such was the intuition of Leibniz, who first suggested the need for such a science, which he too called *geometria situs.*

The appearance of things is a quality we perceive, and it might be generally stated that topology is concerned with perceived shapes, while geometry is concerned only with measured shapes. While geometry is usefully applied in real-world observations, Galileo's tract, *The Surveyor,* identified geometric analysis with the new post-Aristotelian modern science of moving bodies, depending upon Euclid's rational system of arguments from axioms and thence proceeding in the mode of abstract argument. Although Euclidean figures and constructions need not exist anywhere in the world, floating through the emptiness of formal logic, their power for measurement results precisely because they maintain a perfectly abstract and neutral *symbolic distance* from whatever is being measured. By contrast, topology looks at the world and asks where and how things are placed, how they are actually *situated,* what is their *situs,* what pragmatic (and finally aesthetic) consequences follow from placement. Overall, topology asks how shapes relate to placement and position.

The topologist's concern with St. Peter's Basilica in Rome would not be with the size of its great dome, but with the way its shape relates, as shape, to the massive baroque body of the building below it. From that perception he would look outward, for example, to the columns enclosing and thus forming the grand Piazza or perhaps would look inward to Bernini's baroque *baldacchino,* a gorgeous canopy placed over the tomb of St. Peter. Yet in no technical sense would the topologist be concerned with the actual measured size of the basilica—its shaped proportions alone would be the central concern. The four twisted columns and their baroque support of the *baldacchino* would, despite being built according to certain sculpted measures, be significant for their aesthetic torsion. Similar torsions occur in baroque music, where counterpoint increases our sense that shapes of melody are always changing, but also returning to their original contours. At first sight there might appear to be something indeterminate about the way we perceive shaped things and localized places. That is why Leibniz in an almost casual throwaway had mentioned the need for "geometry of position," casually using the word *geometria,* though he was not intending a science of measurement. It was soon revealed just how strange his idea could be.

In the long history of mathematics, topology is a surprisingly modern discipline, nowadays so advanced that it provides the foundation of much of our most important current research into cosmology, terrestrial ecology, computation, neurology, brain function, and numerous other fields; yet topology, unlike relativity, is not a household word. In his book, *In Pursuit of the Unknown,* a study of some famous equations, Professor Ian Stewart says this: "On the whole, you won't run into topology in everyday life . . . but behind the scenes, topology informs the whole of mainstream mathematics, enabling the development of other techniques with more obvious practical uses. This is why mathematicians consider topology to be of vast importance, while the rest of the world has hardly heard of it" (2013, p. 106) Much of advanced science is like this, certainly, but how many Americans can name the makers of the few great American films? It makes no odds that the mass of humanity has never heard of partial differential equations or the Second Law of Thermodynamics, or the Danish director, Carl Dreyer, despite their importance. Yet topology is basic to the programs that enable computer science.

When topology began its career, however, it deliberately avoided measurement of magnitudes, as we shall see, and beyond that refusal of measurement, another equally paradoxical defining trait needs at once to be mentioned. Not only is topology a special way of seeing shapes as unmeasured places, there was yet another idea implicit in the original Leibnizian intuition, which sought to show that the space of Earth is an imagined *quality* of our existence—a lived scene alive with endless displacements.

This central discovery involves a plasticity: *innumerable shapes exist, which preserve their basic form, even when they are bent, twisted, stretched, or otherwise similarly deformed.* If there is an order to the way positions alter the meaning of places, the physical meaning of St. Peter's we might say, there is also a wonderment about the way we understand the stability or instability of shapes and forms. How do we recognize people by their gait, if not by seeing into their bodies? The question had long lain dormant, as to the breadth and variety shape as such might assume, but it seemed to have two cardinal aspects.

In 1735 Leonhard Euler had taken his first step in this direction, when he solved the local puzzle of The Seven Bridges of Königsberg, and with that solution established *analysis situs* in what may be its networking modern mode. But soon, as I have said, he invented what is called his *Polyhedron Theorem* of 1750, and in a way this theorem cuts deeper into the question of changing shapes. Thinking about edges, he discovered why single, individual objects are shaped according to a principle of stable coherence, no matter how different they may appear to the eye. Both discoveries stressed the placement of points in space, including the way objects locate their various internal parts, producing a unity of components to produce what we may call a complete form. This suggests that topologically we are concerned with local character and location as aspects of our living situation on planet Earth, for Earth itself is a body, topologically always remaining a sphere, no matter how irregular or changeable its shape may appear to be in our perceptual experience.

Euler was building his two discoveries on the fact that, despite appearances, although a place may display a changing shape—from a gentle rise to a catastrophic sinkhole, or from a pyramid to a sphere, from a square flowerbed to a round one, from an adolescent body to a mature one—it still maintains a topologically invariant stability of form. For example, a round tower and an octagonal tower share a more basic cylindrical form. The paradox already surfaces, namely that the topology of an initial shape has undergone alteration, but somehow, again *despite appearances,* change of shape has not occurred. Or as we often say, the shape is still more or less the same, even though it is not always easy to say what the words "the same" precisely mean. In common parlance we find ourselves using suffixes such as *-ish* or *-esque,* as with roundish or picturesque. Preromantic thinkers saw this situation coming and would have understood that a landscape might well be picturesque, that is, rather like a picture. In an era like ours, where we are so clever at epitomizing situations and are nevertheless still confused, the *ish* and the *esque* are in dire need of a science like topology, such that sudden transitions may find good order of sequential change. We live in the age of Hamlet's uncertainty and we cannot quarantine our doubts.

Wherever there is a region or a domain to be studied in relation to troubled, shifting boundaries, as in surveying and mapmaking, topological questions may be raised. Currently the mathematics of the field is not only pervasive in science, but it is also highly sophisticated in its entertainment of ambiguity. Consider that Euler's discoveries led to a field described, almost querulously and not entirely without contradiction, by James R. Newman in his monumental four-volume anthology and commentary, *The World of Mathematics,* Vol. I:

> Topology is the geometry of distortion. It deals with fundamentals of geometric properties that are unaffected when we stretch, twist or otherwise change an object's size and shape. It studies linear figures, surfaces or solids; anything from pretzels and knots to networks and maps. . . . Topology seems a queer subject; it delves into strange implausible shapes and its propositions are either childishly obvious (that is, until you try to prove them) or so difficult and abstract that not even a topologist can explain their intuitive meaning. But topology is no queerer than the physical world as we now interpret it. (1956, p. 570)

This may sound like an unstable mode of science or perhaps science fiction, and yet topology is concerned precisely with *continuous* and in that sense stable transformations of the order of things, where fundamental contours remain the same, despite being twisted, stretched, bent, dented, or whatever, *just so long as they are not cut into separate pieces.* This last is the crucial requirement at a certain stage of the mathematics, and if this is true for individual bodies, it is no less true for the body of a city or more abstractly, the body politic. Just like fractures of different bones in the body, the partial cracks or complete breaks (sometimes caused by barriers such as the Berlin Wall), will display the difference between a continuous topological transformation and a decisive cut separating the single limb into a radically different, two-part object, a divided shadow of its former self.

Leonhard Euler Thinks about Bridges

A further possibility now arises: as distinct from transformations occurring within palpably unified things, with animal bodies, hoops and such, there may also be a wider region or externally more widely extended placement, such that more than one object comprises a group or combination of entirely separate parts which *as a group* comprise a single *composite* shape.

This combinatorial coherence is a critical interest of Euler's first discovery, in 1735. We began by asking, what indeed is a shape, in any case? By and large it is easy enough to see the look, the outlines, and sensed forms of objects as identifiable items, which we might perceive as *distinct shapes.* All such objects display shape and seem to be unities, simple enough *things,* but the most important shapes for living beings are compositions of things within things within things. Only then can natural beings encounter the force of things *positioned outside the body,* at which point the individual becomes part of a larger group, which in turn may be said to obey the principles of *analysis situs,* by fitting many parts together organically to create a living or at least stable whole. Anatomy commonly understood is intended to reveal systems of organic interaction, and these manifestly have something to do with hierarchies of living cooperation between and among the connected parts—as anyone knows who has torn a cartilage in the knee.

It was in fact an odd accident that led Leonhard Euler to devise a not dissimilar anatomy of space when, after some professional hesitation, he solved the riddle of the Seven Bridges. In the riddle the Königsberg traffic had to obey a controlling pattern of pathways. The picturesque old city was home to Immanuel Kant, Christian Goldbach, and other intellectuals, though also a town that bored and depressed Stendhal only a little less than the Free City of Danzig (as he tells us in his private *Journal*). The Swiss-born Euler was already a famous mathematician, then resident at the royal court in St. Petersburg, well known to be one of the world's fastest calculators, a dazzling theorist and an immensely prolific contributor to his field, when he was asked by a friend, the Mayor of Danzig, to solve the local riddle in mathematical terms. The assigned task was to find a way to cross all seven bridges of Königsberg—divided by the River Pregel, flowing around a *central island* before draining into the Baltic Sea—but only under a particular requirement: *The task was to cross all seven bridges, but without crossing any bridge twice.* In passing I should say that the island, as place to be crossed, will be an important metaphor all through the following essay.

At first the great man thought this was a trivial joke, but then he noticed that not only was it impervious to "common sense" (his first thought) but that neither geometry nor algebra nor arithmetic could answer the riddle. He found that the city's central island and its bridges comprised a system of topologically constraining vertices, edges, and loops that could be graphed with separate points and lines, something like a wiring diagram, a closed circuit of permitted possibilities. Location of the bridges as a group of pathways became for him a complex unity, a bendable complex mathematical object, whose parts were effectively the moves on a massive chessboard, the City, its bridges, its central island, and the River Pregel.

The length of the bridges and size of Kneiphof Island were not relevant. Size and measurement were no longer the issue for the citizens crossing to work or strolling at their leisure, because the design of the city as a whole, its governing shape as an almost living organism, contained the answer to the puzzle. Therein lay the first of Euler's two great insights, which he described in a 1735 St. Petersburg lecture, subsequently published under the title, "*Solutio problematis ad geometriam situs pertinentis*" (The solution of a problem relating to the geometry of position)—included in Newman's *World of Mathematics.* Euler saw that the bridges needed to be understood in a special way, which we would call a network, where measuring and sizes were irrelevant, and in the following words he introduced his unprecedented solution:

> The branch of geometry that deals with magnitudes has been zealously studied throughout the past, but there is another branch that has been

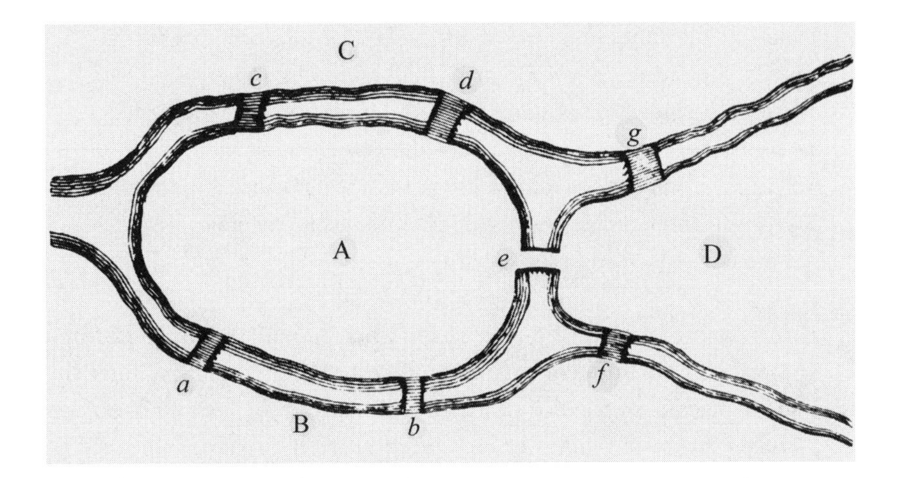

Fig. 1A. Euler's engraved overhead view shows separate landmasses and their bridging connections marked, respectively, with uppercase and lowercase letters: A, B, C, D and *a, b, c, d, and e.* Kneiphof Island is marked A, and it has five bridges. Euler depicted a system of lines or edges crossing various borders. Diagram reprinted from "The Seven Bridges of Königsberg" by Leonhard Euler in *The World of Mathematics: Volume I,* edited by James R. Newman (New York: Simon and Schuster, 1956), p. 573. Originally published in "*Solutio problematis and gemometriam situs pertinentis*" (The solution of a problem relating to the geometry of position) by Leonhard Euler, *Commentarii academiae scientiarum Petropolitanae* 8 (1741), 128–140, Figure I, Plate VIII.

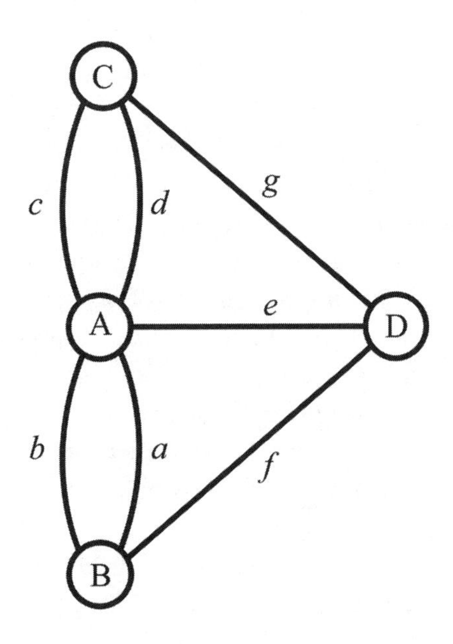

Fig. 1B. Using the same letters as in Figure 1.1A, a modern *Graph* abstractly shows the layout of landmasses and bridges linking them. The number of bridges meeting every landmass may differ: thus C has three links, while A has five links, and the positioning and their odd or even number of linkages determines the possible pathways. Such networking underlies modern computation.

almost unknown up to now; Leibnitz spoke of it first, calling it the "geometry of position" *(geometria situs)*. This branch of geometry deals with relations dependent on position alone, and investigates the properties of position; it does not take magnitudes into consideration, nor does it involve calculation with quantities. But as yet no satisfactory definition has been given of the problems that belong to this geometry of position or of the method to be used in solving them.

A century elapsed before the year 1847, when Johann Benedict Listing's book *Topologie* gave this geometry its modern name of "topology" (literally, the logic of place), "By topology we mean the study of the features of objects . . . *without regard to matters of measure or quantity.*" Again, avoidance of measure and quantity! Without getting rid of the texture and outward forms of things, Listing's new name was much handier than the standard Leibnizian term, geometry of place (*geometria situs*). To reject the immediate goal of measuring and quantifying indeed amounted to a revolution in thought.

Without measurements the method supplied the idea that position implied sequence, which in turn constrained the concept of evolving shapes. Furthermore, we find that we are noticing groups of separate but connected shapes, which together create a web of interlinked positions of those shapes or objects. Ultimately we have formed a picture of multiple interconnection (which we could also do when treating a single object with many parts, such as a living organism). We treat the scene as if it were always undergoing a translation of placement.

In effect, as soon as a complex sequence is treated topologically, the several parts become one single supersite, and whenever such a singular object is multiple in form and meaning, we approximate conditions of much human perception. Indeed, when discussing J. J. Gibson's view of ecological perception, as a matter of edges, we shall see that when shapes are actually combinatorial mixtures that create a paradoxically numerous Oneness, the basic topological approach is necessary. In this vein Donal O'Shea introduces his splendid historical and theoretical account of *Poincaré's Conjecture,* beginning with the technique of assembling whole Atlases out of particular smaller and less global regions, for the Atlas makes a "global" unity of what would otherwise be a disconnected heap of observations and surveys.

If this combinatorial coherence is a fundamental fact of material order, it can only be imagined, and it follows that the function of topology remains the central conceptual discipline of all design. In an imaginative way we are hinting at the artistry of form, hidden inside our numbered universe, which eminently includes our own minds. True, Euler did not provide a rigorous proof of his main principle, though he demonstrated his result, namely that the mayor's demand could not be met—it was not possible to avoid crossing one bridge at least once. The St. Petersburg paper showed that the riddle depended entirely on the number of starting points (the vertices marking each section of the city) directly linked to starting points of each traverse. This use of the principle of "degree," the networked linkage to starting points discernible in a continuous graph of the whole setup, was the essential thing to note. Every traverse had to begin from some point (a "vertex") positioned on a "local" landmass. *Topos* and *locus* were now made a central issue.

Trusting only that his official St. Petersburg paper showed how the bridges involved a topological paradigm, he demonstrated the role of pure sequence in the determination of the wholeness or the fragmentation of any object. Euler had clearly understood, when mentally diagramming his insight, this latter question as to how the bridges might permit a continuous or only a partial connection, but he did not provide a detailed proof. This proof used the line-drawing graph method described by Carl Hierholzer, but not until 1873. It is clear that Euler did not need any other graph of the Bridges except what he imagined—he later in life did much important work in spite of going virtually blind! Yet there was an implicit line drawing of a graph in the mathematician's mind. Furthermore, already by the early eighteenth century common industrial advances might well have provided a sufficient model for such combinatorial imagination, since familiar knowledge of machines showed they were unified yet constructed from bits and pieces. The machine was yet to achieve its hegemony, but such devices were no doubt frightening to many who were accustomed to hand labor, and becoming a slave to the machine early became a threat. In a different direction, there was beginning to be a new openness to the power of harnessed fragments, as we find in the writing of Friedrich Schlegel. Among intellectuals, notably among poets and artists, the conviction often was expressed that we must think critically about monumental statues—those testimonies to vulnerable yet established power—when, in a more natural context of combination, poets had always known that any evolving site could be made to seem object-like, even though multiple in its parts, like the famous *Laocoön* group, just so long as the whole is shown to have an invariant living shape. Oddly enough, in the context of topology's first beginnings, the organic moment of continuous connectivity within the apparently disparate array might almost define Romantic imagination.

That complex evolving connectedness certainly is what Euler discovered, and his discovery leads eventually to modern combinatorial analysis. At first, position and position alone was the issue, and he was able to show that it was impossible for the strollers to cross all seven bridges without repeating at least one traverse. Given the layout of the city, the citizens could not win their chess match. It is as if topology were designed to reflect the archetypal importance of cities, with all their complex social needs. In a fine chapter on the Bridges of Königsberg, David Richeson's book, *Euler's Gem,* shows that the graph theory arising from Euler's first paper has myriad applications today, which include computer science, networking, social structures, transportation systems, and epidemiological modeling. This last field is a graphically understandable case: the spread of an Ebola or MERS virus, for instance, depends entirely on how many contacts the infected victim has had, from which contact-*positions* they have occupied, thence developing the exponential number of bridges of infection their contacts may unwittingly create.

Bridging and connection are close relatives in all such situations, and connectedness is a major principle for topology. Loosely we may think of *connectedness between* and *connectedness within,* and note that Euler first focused on the former, since the bridges allowed citizens to move freely between different parts of the city, locations separated by the River Pregel. In formal terms Euler's Königsburg is a *simply connected space,* which means that its significant places in town are "path-connected," and all the paths across the bridges seek, if possible, to form the whole combined transit into a single essentially uninterrupted unity.

Without getting into the philosophy of Martin Heidegger, one can hardly avoid the thought that narratives also trace pathways. Such is the major method of storytelling in the quintessential late modern author, Jorge Luis Borges, who constructs narrative labyrinths of all sorts, for instance his parody of a spy story, *The Garden of Forking Paths,* or his kabalistic detective story, *Death and the Compass.* In any case, real or fictional, successful or not in this sequential enterprise, the path-connections produce a network of lines and nodal points, a kind of matrix, so that all the pathways belong together as what we may call "lines of connection," from beginning to end. The City then becomes a grid or unification of traffic, permitting Euler to imagine composite patterns, which altogether comprise a complex single site or *situs. Analysis situs* then becomes the testing of connectedness for topology, which develops into many further corners of the question of transforming or deforming shapes.

With such radical insight into the geometry of position, Euler had invented a disciplined account of site, position, and topologically defined sequence, and he changed the world of mathematics. Ian Stewart's bold claim that "topology informs the whole of mainstream mathematics" is scarcely exaggerated. The discovery makes possible an imaginative leap beyond the narrow localization of individual humans, beyond their personal sense of measurable space. Present day acceptance of the unrivaled prestige of science tends to exaggerate the relevance of very large numbers, as found in astrophysics, but we humans live in a much (even cosmically) reduced scale of things. In every respect the quest for unity, whether in thought or in a material universe, either large or small, may for humans be associated with the Bridges of Königsberg. Unlike Euclid's methods of calculating spatial extensions, the analysis of bridge and site was a rethinking of both space and time for a world not unlike our own, a world to which we still belong, although now Kant's city has a Russian name as a result of annexation by Soviet Russia. Euler's solution involved a simple walking motion along the citizens' paths, and in that sense the solution implied that space and time, taken as the abstracted terms of relativity theory, could only be "mere shadows," as Minkowski phrased it. As we get older, time certainly runs away, disappearing with the speed of light, or perhaps, with the speed of thought.

The groups formed by position as treated by Euler and his successors share in a combinatorial process composed like music of a sequence of "tones" or moves altogether producing (or, it might prove, failing to produce) a single composite and yet continuous pathway, a single "movement," as musicians might say, and this perception of movement within the structure is what enabled Euler to understand *situs.* Constraints of relative positions (as if one heard cadences sung in chess moves) finally lead to a whole new set of impulses, constraints, and invariants of movement. The Bridges were dividing edges, and they allowed the site to become the living situation. Only in this sense did the seven bridges measure the vanishing flow, which was time's commentary on the passing citizens and their friends. After millennia of geometric measuring, this teasing puzzle showed a secret pattern hidden beneath a surface manifold—a How buried deep inside a Why.

As the artist knows, in aesthetic perception the sizes of things simply disappear into a deep background of order and are not part of the shaping experience. What then are the shapes of the growing leaf? Or of the embryo? They are part of a measurable world, in one sense, but we speak of shape because we need, by design, a less exacting approach to the things of life . . . or so it seems to me, and so also to the artist. For shape is not perceptually conveyed as measurement, although in the 1940s beauty queens were known for their dimensions. Only those who have no sense of shape will need to use a tape measure in the interests of beauty, as only a scientist, someone like Galileo, will require the use of microscopes or telescopes to think of looking beyond. Primitive it may seem, but to know a shape is like following a winding path or blindly recognizing a three-dimensional body by touch alone. To avoid our being influenced by conscious or half-conscious thoughts, we use instruments to limit error. For example, the path of a particle may need a microscopic device to be registered initially, after a collision; but for sensible processing, the picture we observers get finally must be made available to us on a macroscopic level. Mathematics may reduce aspects of the particle's movements or its spin to an equation, but that reduction to number is not significantly its "shape." Even a quantum mechanical wave is not so much a shape as it is a statistical estimate of a continuously changing electromagnetic charge. Then too, on a macroscopic level shape is also global, since it is perceived "in the round" or is imagined as an object having a three-dimensional form. A shapely design is never an abstract algorithm, though a calculation of shape can be made "after the fact," as it were. A dolphin and a porpoise have different shapes, allowing us to identify each species, but we do not need to measure them to see their difference.

Treated under a correct formal theory, the many indeed can become one. We may say that shapes identify bodies, while measurement identifies distance, which in turn may allow us to think of bodies proportionally, because comparing measurements of individual parts of the whole yields the central notion of the Scale of Things. It is no accident that Poincaré insisted on questioning our knowledge of topological basics, asking, for example, why we "suppose that we know what is the manifold of points of space or even one point of space." There is nothing conventional about such questions, and they lead to a properly wondering attitude. He also asks what we mean when we say that we are making a cut, since a dividing of a region of space can well be an extended surface, as described in "The Nature of Space," a lecture on *The Value of Science.*

Here Poincaré is reaching out to the most fundamental idea of divisions of site and situation. Only the zero dimensional single point, so mysterious in itself, cannot be further divided—a terminus of the act of cutting (*coupure*). Yet we can also reach beyond mental extensions of the human body and its physical contacts. Suppose that we are considering the larger scales of analysis, whether in human affairs or at the extremes of large and small material things, for example, matters of political economy or climate change, then the serious questions and answers all seem to be of a combinatorial nature, which must be understood as the positioning of the points of relationship. Constraints upon what we find we can actually do in this world are all dependent upon relative positions from which we start and through which we must pass. At the point of human use, of ethics, my essay will return to this question of scale, where shape, place, and measure collide, if they do not exactly merge.

For the artist, shape itself turns out to be an elusive imaginative concept, a fantastic problem almost, as, for example, a painter tries to convey the look of a hawk's beak. For him the issue is always a "meditation on a hobby horse," to use Ernst Gombrich's title for a lecture on aesthetics. How can a line drawing convey three-dimensional depth? How do the wooden body and stick legs conjure up the living horse? Similarly, for the topological mathematician, things are not always materially "filled out." This mathematics is not like an earlier more transparent Euclidean geometry, which developed, as one cynic recently said, to advance the cause of ancient real restate; topology is more like music or sculpture, evoking the touch and memory of familiar bodies foreseen before the melody is ended. These always reflect temporal flow, while with a quite different feel, much of our daily experience cuts our lives into divided flows of experience, interrupted artificially by clocked appointments to meet. Because we live in a disturbingly fragmented even if "global" world, the topological treatment of cuts and interrupting edges may eventually be seen as one of its greatest contributions to thought. A true sense of edge and its relation to bending, stretching, and twisting in fact permits a reasonable method of gaining continuity.

In the interests of this continuity we must turn to the question of metamorphosis as an experienced phenomenon. An accomplished commentator on physics and mathematics, Michael Guillen, in his book *Bridges to Infinity,* praises the poetic refusal of narrowly conventional thought: "Some artists who look upon an object tend to see all the different ways they might depict the object and still communicate its essential being," instancing Monet's paintings of Rouen Cathedral, which restore a fundamental strangeness in the shifting perception of objects (1983, p. 156). Their apparent permanence in change is deliberately disturbed, even as it is being painted and virtually fragmented by the artist. Nor must we advance into the most recent period of art history to discover such plays of formal torsion; they are to be found throughout the history of art, and not only in the visual arts—no doubt because artists discover intuitively what Euler had to discover mathematically—the persistence of a fundamental type in the midst of changing shape, perhaps a mathematical analogue of Goethe's archetypal *Ur-pflanze.* Later, virtually all Darwinian observations in the field are attempts to define the topology of changing, emerging bioforms. We know that far more movement occurs in the universe than common sense perceives, but strikingly in the biological world of embryology—as we will later see in remarks on Thom, Deleuze, and folded edges—the plasticity of shaping the growth process is inherently, as Bradley Bassler wrote me in a personal note to me, "more topological, closer to the underlying imprinting shape out of which the fully grown organism emerges." Embryology may be the best case, but crystallography, like evolving musical variation, also exemplifies such growth, where we may least expect it.

This chapter began with some basic definitions of topology, to which we must return by means of what amounts to a visual cliché. The most famous examples of topological invariance are found in solid objects and mere things, whose familiar knowledge may be commonly illustrated by molding a soft clay donut shape (technically, a torus) into the form of a coffee mug; this process can be reversed, when the clay is remolded into the original donut shape. The apparent form and the intended purpose, holding a liquid and making something that looks edible, are altered, but an originating form is preserved. The essential permanent feature is the hole in the donut, which continues unchanged, even when shifted to the handle outside of the mug, meanwhile preserving the essential continuity of all else in the object. The preservation of an immanent shape is a fine example of what mathematicians call *invariance.* We might think otherwise, but a closer look tells us that the hole in the donut does not cut the object in two nor fragment the coffee mug; the hole is the topologically defining feature of both donut and mug, and they are therefore *homeomorphic.* If we make objects out of soft materials, we quickly see the range of possible objects undergoing the topologist's permanence in change, as for example if we were to reshape a cube of putty into a round ball or a dish or a pyramid—or indeed a host of other familiar shapes. Something about these shapes refuses to give up its invariance.

My interest in such transformations is not, however, instrumentally material, for consciousness in its imaginative role cannot be a material state; if it were, as Locke and others noted long ago, we would all feel pain and pleasure in distinct, exactly identical ways. When thinkers of any kind, when perceivers see something, emoters emote, analyzers analyze anything, parts of the brain may lie dormant while other parts are busy sending signals to each other, and all this biochemical activity can be analyzed by neurophysiologists in a material fashion, by studying brain functions on the most delicate basis, but the experiences and the consciousness as such are not actually touched by such analysis—for metaphysical reasons. Fortunately, the notion of shape does not require narrowing its range to external objects like tables and chairs and carbon atoms. Rather, while the idea of shape allows us to add to our endless list of things and objects, we can move over to an equally long list of complex situations which we might well name objective conditions or interpersonal discussion, such as might occur in ordinary social life. Our ideas and thoughtful inventions have a long reach of association, relating to a boundless set of variations on the theme of life, as if we were all the children of J. S. Bach and, as Douglas Hofstader has suggested, life is an endless braided art of the fugue, a system of contrapuntal analogue-forms.

When we humans actually control the external workings of Nature, as in the agribusiness, our hard task is to respect the Song of the Earth. Vladimir Vernadsky, the Soviet ecologist, would say that we cannot avoid our responsibilities to the *noösphere*, nature cooperating with mind, an Inmixing with an Otherness, to quote Lacan, and our metaphors and other figures of speech reveal this mental aspect of our relation to Nature and our desire to exercise control over Nature. Timothy Morton's "ecology without nature" is paradoxically a natural idea, in that its negative is produced by its author's natural mind (2007). Nor, on the other side, should we avoid the constraints imposed on us by artificial or material forces and conditions, for our sudden creative insights come to us in the form of figured expressions, which we ourselves think into being, mainly as metaphors whose origin and dream come from the world of fact. The strong connection in the books of Etienne Klein, especially his *Chronos* and his *Conversations with the Sphinx,* between science and literature amounts to an admission that the Goddess Nature is nothing, if not a melancholy wit.

Despite the power of our neutralizing numbers, to present space and time as if they were measurable quantifying abstractions existing all by themselves in a vacuum of logical purity seems somehow inadequate. They involve *dimensions of something, surely.* A human consciousness mediates their mathematical utility, and pushes these dimensions over into an actual shaping of actual things. The human component of thought demands that we think first about what we might call social topology, since we are linking disparate visions, of artist, scientist, seer, and citizen. With due respect, then, for the ambiguities attending all analogies, let us begin as close as we can to our own consciousness and its world. The Bridges and the City were manifestly material situations, things, objects, concrete arrangements, but as I have insisted, their form had first to be thought, if they were to be made real and humanly present. The Königsberg path-connected solution has many analogues, all of them important for complex thought, which universally involves relations *between* things.

The Dramatic Analogue

Consider then what might at first sight appear a remote parallel, but one I regard as quintessential—the drama and its concern with logic. Plays and theater almost perfectly define the imaginative drama of ideas as forms. Their performance embodies our thinking as an experience, an imagined embodiment presented through persons *acting,* which provides the topological equivalent of focused consciousness. Let us then take a familiar case, the play *Hamlet.* In general the drama, as distinct from glitzy production-value show business, is a testing procedure for the logic of mental flexibility, whenever the play presents deformations in and of the social order. A good play is logic embodied, but plays thrive on instability of situation. Alone on the ancient or modern stage, alone on the medieval *plateia,* the actors *en-act* our human *re-actions* to the forces of stress, the forces of bending, twisting, and stretching social position, social authority.

The identity of each person we call a character shares in a corporeal principle of invariance and, as the Renaissance scholar and critical theorist Harry Berger, Jr. shows in his book, *Fictions of the Pose* (1994), an actor enacting a pose struck by a character he is representing on stage is in fact giving us the picture of a picture of a picture. This Russellian, inherently self-reflexive paradox haunts the whole modern universe of portraiture, including photography and its snapshots of persons. How then do we stage the already staged, the already posed? Is a pose in any sense a "real" shape? To illustrate the question, let us turn to the most celebrated dramatic exploration of stable / unstable character, the young Danish Prince for whom a question of permanence arises with the first words of the tragedy, establishing the primacy of question itself. We question whatever is uncertain, especially an ambiguous, not obviously real presence, an appearance that may in a moment dis-appear.

The uncanny opening scene of *Hamlet* thus issues, as it were, from nowhere, when a voice calls out two words: "Who's there?" and before we can take a breath a spectre, the unidentified ghost of Prince Hamlet's father, appears to the guards and the Prince's friend, Horatio. What is this shape? In terror they ask if this ghost is only an illusion; they call it an "apparition," a "thing." The ghostly monarch is actually a talking shape, and his dramatic appearance is one of stark, catastrophic collision with common reality, until his son the Prince tests the apparition for its material knowledge of fact. Yet later in the play the Ghost comes and goes with indeterminate outlines and clouded solidity. What then is constant in the King's appearance? Hamlet the father is in some sense Hamlet the son. Returned or rather escaped alive from his sea voyage, the young Prince later cries out: "'Tis I, Hamlet the Dane," at which point he has become his own father. Using the language of identity, the play studies the very idea of shape and its changing permanence.

Shortly after the opening scene ends, Hamlet himself utters the key word, "seems." To his troubled mother he replies: "Seems, Madam, I know not seems." I mention this term, rich with meaning, because it demonstrates an important topological feature—the transformation of forms needs to be somehow part of a *continuous* flow, and in Scene Two the theme of seeming flows unobtrusively out of the Scene One theme of the ghostly apparition. The second theme finds its continuance, its more general connections, in the second scene. Anticipating the question of being which Hamlet later raises in his most famous soliloquy, the play begins by asking how we should understand the vague realism of our observations, indeed how we should decide on the boundaries of any object, given that every object has to some degree an ideal framework of interpretation.

To use the later words of the Prince's famous soliloquy, there is a testing of the meaning of "this too, too solid flesh," as if the play were anticipating Leonhard Euler's term, "solid angles," when analyzing the famed five Platonic solids. There is a constant testing of solidity (in political respects also) through the analytic functions of drama and staging and, above all, of the actors manifestly acting, of taking action, and of course drama is a fine example of testing ideas which inter-act. As the history of the discovery of topology shows, and we will later discuss, the question may be asked: how spectral is any apparently single thing? What if a single object is actually an odd combination of smaller objects? What happens when shapes are actually compound? What if the King in *Hamlet* is actually more than one person? In Renaissance thought, in fact, there was a common political myth that gave the king at least two bodies, since he was a single person, but also the whole body politic incarnate. Perhaps all persons embody a similar multiple selfhood, which nonetheless is identifiable in just that one particular person, for as it turns out, in theory the combinatorial play of bodies and objects was the epochal insight inaugurating topological analysis.

A less metaphysical example might better bring the Prince's mystery of local identity down to earth. Ron Howard's docudrama film, *Apollo 13*, perfectly illustrates just such a moment of illumination. In 1970 an explosion caused inexplicable major damage to the spaceship, forcing the actual *Apollo 13* crew to abort their mission and to attempt an emergency return to earth. When the explosion occurred, Jim Lovell, the captain of the spaceship, radioed back to Mission Control, "Houston, we have a problem here." Let us change the script for argument's sake, translating the emergency message into another commonplace American idiom: "Houston, we have a situation." "Situation" would have been harder to hear than the word "problem," but it would have been more precise, for as it turned out, the astronauts' situation was not just a problem; it was a topological puzzle, solved only when the crew pieced together a unified, functioning substitute for their original fully functioning air supply, which meant they had to reconnect devices for "scrubbing" away the excess of CO_2 that must come from exhaled breath. Too much exhaled CO_2 would simply have killed the three men.

With help from Houston Control they assembled a homemade combination of elements, close enough to the original sequencing in the original equipment, thus producing a topological permanence in change. They achieved a new "situation" but one bearing sufficient topological similarity to the previous one. Unlike the original equipment that was designed and built in factories on the ground, the homemade solution was not a result of perfectly engineered joints between the parts, but by duplicating the placement, sequence, and positions of each substitute transition, the astronauts and their Capcom guides found a way to create a coherent pathway to save their diminishing supply of oxygen. In fact the task was finally to fit a round peg into a square "scrubber" hole. Altogether the solution to the disastrous problem was a combinatorial agglomeration of several parts permitting a continuous flow, and to see this sequencing of positions on the pathway was to think topologically. Famously, at the critical endpoint the separate required parts were "glued" together with strips of duct tape, which permitted a continued flow of CO_2, where the chief difficulty—for engineers on the ground and then in the spaceship—was precisely to achieve a topologically continuous pathway, virtually in the manner Euler first analyzed.

Topologists indeed customarily speak of "gluing," albeit on an sbstract plane, when they follow the deformations of one shape changing into another. What I. A. Richards called "metaphorical disparity action" requires that when we uses figures such as analogy, simile, and metaphor, and even the more nominalist metonomy, we are in fact gluing together disparate aspects of our conscious awareness. Cohering conceptual forms and coherence in Nature are thus never in principle quite as perfectly flowing as we may wish; they are not so distinct from combining rolls of adhesive tape, whose continuous edges, rolled and then unrolled, "inmix" deeper variety and conflict among separate parts. As our visionary poet, Walt Whitman, several times wrote, the basic life-principle is "adhesion." He would have admired the calmness of the astronauts, and he would have recognized a truth in my analogy, whereby the space age story is less fanciful than might at first appear, picturing, as it does, a larger design of obscure, vague, or even uncanny conditions, the conditions of living motion. These ambiguities topology is able to clarify, with its maps of exploration, or rather, its plans for unexpected combination.

Continuous motion is not always a sign of life, however, and here a window beyond the space age stands open before us. With Apollo 13 the use of the adhesive plastic tape looks different from different distances. Seen up close the tape is one part of an artificial, home-made engineering substitute for another pathway for CO_2 to flow through. But at a distance, whether in thought or fact, the tape belongs to a membrane whose bending (Gell-Mann would call it the "plectics" of the arrangement) is part of an organic life-system. The broad suggestion is that when topological is fully operating there is no clear line between what is alive and what is not. Between these two different states of being, between their parts and the whole, there is no clear boundary; there is only an ambiguous threshold whose edge dilates and contracts in continuous motion. For the astronauts a successful crossing finally meant life—or rather, life's version of success. On another view, admittedly, a biological machine may be the most desperate, if understandable wish-fulfillment.

Handling human catastrophes is not unlike handling sudden success. As soon as a complex sequence is treated topologically, the several parts become one single site, and whenever a real, singular object is complex in meaning, we have the same possibilities of intelligent perception. If every site has to be a palpable thing, a pyramid or a persimmon that could be continuously deformed, this would still be a discrete entity or at least a "mathematical object" based on singularity in the ordinary sense. Yet we have equally seen that when shapes are actually strange compounds, like the King's ghost, a sort of body politic, a numerous unity, the same topological approach is useful.

We can go further, asking: what if all bodies and all singular things are always combinations of interacting, smaller, equally individual things, and therefore *singulars* always constitute a site as a *dramatistic* situation, to think of drama in Kenneth Burke's sense, as originally developed in his *Grammar of Motives*? Such a question would apply, with equal and more consequential force, to interactions between actual human persons and their decisions. Yet if combinatorial coherence is a fundamental fact of material orders, it must, as with human persons, be imagined, so that topology remains a central conceptual discipline of all design, and "designs," because design is shape subject to phenomenological limits. In an imaginative way we are hinting at the artistry of form, hidden inside our numbered universe, which eminently includes our own minds. As soon as a complex sequence is treated topologically, the several parts come together to produce one single complex site, even of disaster, and whenever a real, singular object is complex in meaning, we need intelligent perception.

It is as if the edges of the adhesive tape are the organic prerequisite of the possibility of reslity in its Heraclitan flux, When shapes are actually strange compounds, a weird contradictory object, be it the King's ghost, a clanking immaterial suit of armor, a body politic, or some other numerous "thing," we need a topological approach to pierce the veils of constantly changing appearance. It is as if edges (and hence shapes) order appearance in one fashion, and measurement another, while the two modes have intricate links with each other. In their different approaches they reason about two ways by which the One and the Many encounter each other at some border. The theory of numbers, along with a need to observe the real, is doubtless our surest guide in these puzzles, promising an economy of thought, but on the other hand we humans are fully aware that sometimes, without counting, we may approach ontology, some vastly complex ambience "coming together" to produce a singleness of effect, like the weather or a climate change, which in this context of topological terms one might well name "a situation."

The Arts seem to exist precisely to loosen our preconceptions regarding site, for they ask how consistent these actually are, given the weaknesses as well as strengths of human cognition. In a complex technological era such as the present, it will be perhaps easier to think of intricate networks of communications as more open to frequent mutability of form, than to imagine loaves of bread and apples undergoing physical metamorphosis or perceptual *anamorphoses.* The artist is at home with this distorting coherence, often resorting to the denial of consistency. Meanwhile, in the war between coherence and consistency, topology might be called a game of testing our fixed ideas, whether distinctly hard-edged and consistent, or merely coherent.

About shapes blurred at the edges like apparitions or ghosts disturbing our sense of familiar place, we can only trust our unclear intuition, staring into the dark. The forlorn Lady Macbeth cried out, "Is this a dagger that I see before my eyes?" Yet normal material reality is marked by an opposite sense of things—with the normal we trust our senses to confirm the existence of a thing. We naturally think of bridges and riverbanks as palpable, identifiable, real things. Apples and oranges, mangoes and peaches differ in many ways, but their shapes are not radically different; they do not resemble lettuce or spinach. They are all more or less spherical, and yet if with Cézanne we move them around on a table, their mere appearance to the eye might deceive us, as to which is which. It is true that we recognize shape as a function of outline, of boundary, but when color and surface texture are added to the mix, our perceptions involve a richer, more complex response to shape, mainly because we can now see objects in the round, as more than two-dimensional outlines. We can even imagine the other side, the occluded back side, and as with other aspects of sensation and perception, we benefit from higher orders of thought, whereby the mind recognizes a perceptual constancy—the locust tree of one species is "not quite the same" as a locust tree of another species, and this perceptual constancy will trigger a strong response for the trained, but also the attentive untrained, eye.

Mind as Ecosystem

Topology asks us to seek as much invariant clarity as possible. In the present context fruits and trees and indeed all differentiable objects have their distinct identity as things possessing different shapes, because normal "perceptual systems," to use James J. Gibson's phrase for the function of the senses, allow us instantly to distinguish between trees and telephone poles. Yet scarred wooden telephone poles retain some of the woody texture of their material, their relation to trees from which they are made, and in such fashion outline is a complex variable, even though quite distinctly the tree is not the telephone pole to which it gave birth. Gibson's ecological theory of perception held that our senses pick up all sorts of topological information about what is "out there"—the information exists before us, facing us, in nature. To show clearly what such perception involves, the topologist analyzes shape in the material world and finds that because the fundamental forms of things may remain unchanged even though the perceived object has undergone an apparent change, topology provides a preliminary account of the conditions of perception in the most general sense. This is a broad claim, but it seems fair, and it crosses into domains far wider than logging and the lumber trade.

Thus we must develop a virtually anamorphic approach to perceiving the world, on ecological lines. Bridging between positions—crossing over and beyond—is the main issue in perception as well as cognition, but it is not always easy for common sense to use this crossing technique, at least in mathematical terms. For example, we usually think that our planet is a fairly round solid ball, stretching as Dr. Johnson's poem on the vanity of human wishes curiously has it, "from China to Peru," and we forget that the poetic sense of The Globe, made famous as the theater used by Shakespeare's company, went along with Renaissance explorations and better maps which projected roundness onto flat surfaces known as charts. From this cartographic procedure we begin to see that we model our round world onto a plane surface, by conceiving it as what a topologist calls a *manifold,* in this case a two-dimensional manifold. The projection traces the three-dimensional shape of the planet onto a "flat map"—a phrase from the metaphysical poet, John Donne, used by him to convey being laid out flat, when he is dying. As Donal O'Shea states, "a two-dimensional manifold or surface is a mathematical object all areas of which can be represented on some map on a sheet of paper. (2007, p. 22). He reminds us of one crucial fact, namely that such mapped surfaces are not only mathematical objects, but are also "idealizations of physical reality." This power to imagine shape in abstracted forms, O'Shea observes, is one reason the topologist gains precision in the projection of surfaces and other basic properties of space and place. Ideally one should even be able to imagine two perfect invisible spheres joining each other through a coincidence of points on their surfaces, an impossible union outside the topology of spatial imagination.

Historically, for Leibniz space was not the absolute Newtonian emptiness of "God's sensorium," an emptiness hospitable to gravitational force and the exceedingly complex mathematics of orbital motion. Relative positions for Leibniz gave conceptual power to mere sequence—positions placed and deposited in the most radical fashion, placed here, and here, and here, so that, as we shall see, these sequences could finally lead to a whole new mathematics of dividing edges. Position and edge create consequent wholes, thus permitting what begins as an array of differences to achieve a final singleness of encompassing shape, with site becoming situation.

After millennia of geometric measuring, this was a revolutionary Leibnizian move toward the artistry of form structuring our numbered universe. Michel Serres, the lifelong student of Leibnizian mathematics and their implications, has shown how Leibniz introduced the combination of science and poetry, or more generally, science and imagination, so that, for example, he shows how the smoke and steam of the seventeenth century create the new climatology of modern industry, while animating the paintings of Turner or, we might add, the atmosphere of so many scenes in Dickens. The atmosphere becomes our sphere, in those crossovers. In part because our world is thermodynamically fueled and controlled, we are strangely governed by mixed combinations of material and metaphysical reality, and we are still encountering the mixtures in the meanings of Maxwell's Equations, which led to the twin Einsteinian theories of relativity. Quantum theory and its calculations now begin to illuminate the operations of molecules and cells in the brain. Perhaps the mystery of radiation is another word for thought, but if so, only in the Miltonic sense of his Invocation to Light, which begins with the immortal words, "Hail, Holy Light," a light illuminating the infinitely varied shapes surrounding us.

In the end we ask about the structure of thought and about whatever intuition lights up the human brain, given that it contains well over eighty billion neurons, arranged in extremely complicated networks, all of which give us at least the illusion of unified vision. As the poet Milton wrote, the mind "is its own place," and in this light we might consider a recent scientific debate over the mechanisms of thought, since nothing could be more complex than the biochemistry involved. In conversation with Jean-Pierre Changeux, the distinguished neurobiologist, the Fields Medalist Alain Connes adopts the mathematician's perspective: "We've been supposing that the structure and function of the brain is characterized by a large degree of diversity, but also by a certain invariance, across individuals. *Topology is just the right framework for understanding this kind of phenomenon, because the same topological object can assume many different realizations*" (1995, p. 134; emphasis added). As topological realities, the invariants provide unusual formal stability to an elusive, indeed otherwise mysterious, process of thinking about space and time. To think about these two aspects of dimension is to look into the mirror of mind, and in a primal way it seems to be an early process of growth. Living creatures need stabilizing balance from the very start of life, for they all begin life by crossing an edge, as a fertilized egg or the embryo starts out by changing and yet preserving the topology of an evolving *urform.*

Returning indirectly to the Enlightenment question about what lasts in traditions when things and customs change, Michael Guillen's *Bridges to Infinity* in effect is a comment upon what we need to see, namely what it means to conserve. Echoing other topologists, Guillen recalls the classic definition: "topology itself is the geometry that is concerned with those properties of a thing that are not destroyed through bending, stretching, and twisting, the three specific elements of topological transformation" (1983, p. 154). This principle appears, for example, in a simple metal hoop. Three points marked 1, 2, 3 on the hoop will retain their same relative positions on the perimeter; no matter how much we bend, stretch, or twist the circle, we say that the points are always holding an invariant relationship to each other, so that "2" is always in the middle, "1" always first, and "3" always last in the sequence. The hoop is Guillen's graphic example, but finally many other similar shapes will suffice; the topologist is not surprised that eggs made of plastic can be modeled into cubes or spheres or pyramids, or frying pans, or prosthetic limbs.

By the end of the nineteenth century Henri Poincaré had moved far from qualitative shapes into point-relationships, and thus he formulated Algebraic Topology, such that the original Leibnizian non-measuring criterion could be abandoned, to some degree. Topology could now be made instrumental for measurement, such as industry requires. Precision instruments require exact boundaries, edges, and areas calling for tight control. Control itself appears to be a tricky topological demand. I myself recall from childhood one crude instance of such controlling devices, and today this memory resounds for me with ominous meaning, as I look back on the years between 1939 and 1945, or rather, long before and long after. The memory is that of cattle-guarding boundaries formed with barbed wire fencing that could be stretched and bent, where as children we would climb through them and then let the fence-wire return more or less to its original taut horizontal line. There was already something vaguely dangerous about those fences, though we learned to handle them. The wire is open to an odd, symmetrically reversible sort of bending motion, whose formal principle needs to be understood to avoid unwanted cuts. In his book, *Barbed Wire: An Ecology of Modernity,* the mathematician Reviel Netz has shown how such wire has been used by ranchers and farmers, but also for prisons and the most terrifying wars. Only a Heidegger could write calmly about this method designed to control free movement of both animals and humans, where once again the issue is topological—the barbed wire changes habitable place to imprisoning space, from place to desert, from a home to a political state. The open and the enclosed are divided by an invisible edge, one thinks, but it is always potentially barbed, and whatever the medium or intended purpose, the science requires that we look for permanence in change—the place or soul within, so valued in quality and yet so elusive.

On one view this value can appear as a kind of nostalgia for a lost home, which is partly the interest explored by Professor Jeff Malpas, in his *Heidegger's Topology* and other volumes. Dwelling is a key idea for Heidegger, and in his later work it looks back to a Pre-Socratic archaism as well as to Hölderlin and German Romanticism. Having myself been moved by the poetry of Friedrich Hölderlin and notably the bleak Georg Trakl, and above all by fragments of the Pre-Socratics, I am yet obliged (like Malpas also) to mention Heidegger's unrepentant support of the

Nazis and his manifest anti-Semitism, but one can ask a more neutral question: is the nostalgic sense of place, in itself, a dangerously comfortable sentimental attachment? In this ensuing essay, therefore, I mainly think of place, space, and topology in a neutral way, as if no feelings were attached to their study, but my reader may feel that my perspective on cherished place might be thought vaguely Heideggerian. A dwelling familiar over time is bound to evoke feelings. Like any deep, one might even say infantile, source of an emotion, the sense of place needs to be *seen* as neutrally as possible, yet in societal terms without losing its human touch. Let me simply mention that my version of topology would be less Romantic in origin and purpose than any fixation on place as home, for I am unhappy though moved by Germanic *Sehnsucht,* which of course Heidegger, deviously clever in such matters, was latterly quick to criticize as a lapse of high intelligence and what he called "thinking." Like other children of Pietism he was an expert at recantation.

Such cautionary notes always sound last when making easy later historical judgments, but beyond that, thinking pragmatically, we see that continuity with past customs cannot now comfortably fit our changing world. (My interest in Giambattista Vico issues from such problems, as the reader will see.) If something goes radically wrong with any requirement of continuity and connection, as for instance in constructing a modern building, a skyscraper for example, the empowering shape-invariance with its capacity to handle changing conditions, such as high winds, has somehow got lost. Always the trick involves a search for a basic formal integrity in the object, whereby various principles assure that "deforming" that object does not alter its essential shape. We always remember that if an unfilled donut is soft enough, one can reshape it into a coffee cup, because both cup and donut are complete forms, each having a single hole in their form. Interestingly, one cannot dunk the coffee cup into the donut, which proves the main point. These familiar examples are now graphically available for inspection on the Internet, where every name or fact seems able to dunk into every other one! One intuits the transformational power of *the Net* (and now *the Cloud*) without trouble, in any case. In the words of the distinguished topologist, Richard Courant, such invariant metamorphoses are "in a sense the deepest and most fundamental of all geometrical properties, since they persist under the most drastic changes of shape" (Newman, *The World of Mathematics,* Vol. I, 587). This depth seems an appropriate idea for an indwelling quality of constant inner shape, not so different in feel from the depth spoken of by mathematicians when describing certain equations, or philosophers when contemplating metaphysics, or poets studying the masters.

The intuition of topological invariance, if applied to human experience, which is my chief concern in this book, is that if shapes are to maintain a permanence in change, they must be experienced as aspects of metamorphosis, on the one hand, but also aspects of fixity and even "solidity" on the other. In the spirit of Montaigne's late writing, this paradox of modeling applies to all our human uncertainty. The *Essays* repeatedly show how we humans tape together all the separate parts and moments of our lives, to make a kind of continuity. Montaigne is a model of creative skepticism for our modern world, and as a man experienced in practical life, in all manner of fragmentations, he expresses that skeptical approach by saying that if he could make decisions, he would not write essays.

Most important of all cautions in such conjectures, the critic will see, is that I am moving through and around in different intellectual disciplines, and only centrally that of topology. If the non-mathematical person like myself carefully follows Donal O'Shea, George Szpiro, Ian Stewart, and others who have described the *Poincaré Conjecture,* the precision and complexity involved will be immediately apparent. Those are qualities of all intellectual work and should be allied with the powerful use of analogies, for only with the proper use of analogy (whose soul is metaphor) can we intuit how to deal with large-scale emergents or essentially complex local action and events of any kind. We need a better grasp of disparity in figures of thought than has been available from seemingly authoritative sources. We need to explore analogical principles of likeness and unlikeness, as Paul Bartha in *The Stanford Encyclopedia of Philosophy* has shown in an article on "Analogy and Analogical Thinking." *Scientific American* recently sponsored a large competitive collection of approaches to analogy in science, and the resulting articles on this heuristic are impressive. Our aim becomes exploratory, for it may be that the flow of any force or energy, such as the flow of heat, even when abstracted to a mathematical formulation, is a topological deformation of shape.

One would like to be able to say that we can structurally and topologically relate heat flows and their environmental effects to the topology of the *Conjecture*. Hamilton's and then Perelman's mathematical use of the Ricci Flow Equations were critical to solving the famous conjecture, but we need to think about Ricci analogically, because he was formalizing changing curvatures of space in his theory of tensors, which Einstein later was able to use at critical junctures of his General Theory of Relativity. What is the analogy to the mathematics here, if we are trying to connect abstract and material thinking? In the nineteenth century Gregorio Ricci-Curbastro was working with mathematical analogues to the flow of heat, which in turn lead us to think about our diurnal experience of light from the Sun, and finally in consequence could lead us to think more *clearly* and yet broadly about the life-process. With Ricci Flow there is a meeting of mathematics and nature, with the numbers and the heat flow, which suggests the analogical play of mind. This means thinking in an impure way, to be sure. Yet heat exchanges are critical in these analogues. Heat, energy, and work were the original concern for Ricci-Curbastro, and their combination requires a seriously playful imaginative account. When Richard Hamilton went back to the Ricci Flow equations, he was thinking analogically. In a phrase of I. A. Richards (from the title of one of his later books), analogy seems to me a necessary *speculative instrument,* at least for the early stages of conjectural thought, where we are trying to conceive and then observe the elements of our more elusive human puzzles.

We shall soon be noting that topology thinks in edges and connections when it analyzes permanence in spatial and other changes. When compact shapes like pyramids and spheres become cubes, they also maintain a permanent *connectedness.* Furthermore, in this fashion topology reveals the true nature of analogical thought, governed by a matrix of likeness and unlikeness in the perception of the logic of forms, when these abstract forms undergo a phase-change into actual, real, concrete shapes—into Whitehead's *concrescences.* Figuratively, if these changing shapes were human, we would say they preserve their invariant "me myself," the inward personhood celebrated by Walt Whitman in his first triumph, *Song of Myself,* the subtle metamorphosis in fifty-two weeks of the year that opens his *Leaves of Grass,* the exfoliating American epic.

Unlike other lesser spokesmen—think of pious Polonius advising his son in *Hamlet,* or Lord Chesterfield's letters to his son—Whitman in the first great chant, and then through expanding editions appearing until 1881, actually probed the contradictions of self. In every situation he raised the possibility that each person is a kind of human island, separate but conjoined with the mass of humanity, a sparkle from the knife-sharpener's wheel, born to flame for that brief life while joined in the stream of light.

The stream fascinates us, for it is continuous, which tells us something of value about the island self. We are each of us like the other, but we also live in the conjoint relation to a larger body. In that sense we are figures of the whole species, we are its metaphors—disparate, yet as Whitman always said, so much as one. The noiseless patient spider launching filament after filament, like the poet, reaches across emptiness to another bridgehead, another island of repose:

> And you O my soul where you stand,
> Surrounded, detached, in measureless oceans of space,
> Ceaselessly musing, venturing, throwing, seeking the spheres to connect
>   them,
> Till the bridge you need be form'd, till the ductile anchor hold,
> Till the gossamer thread you fling catch somewhere, O my soul.

The assumption of an active soul makes no pretense to absolute certainty, but the exploring it seeks is based on a natural instinct, which searches continually for that certainty. The topological interest of a poem like Whitman's, or a similar thought-experiment in science, is relational—it inheres in the question every island raises: from what fixed body of ground does the thrust of thought launch forth? Does the apparent separation of the "stand" from the new resting place imply that connections are never certain or perhaps possible, even when the stand tells us we have a safer containing home-base, a continent from which the island leaps away? We have to imagine the possibility of a larger pathway or place in which to move, without utter discontinuity. We have to imagine Ariadne's thread, an ancient mythic figure for successful action and return from dangerous distance. In *Speculative Instruments,* I. A. Richards included an essay in which he recalled that Shakespeare's *Troilus and Cressida* finds a rich double meaning, when he conflates *Ariadne*—Theseus's savior from the monster and his labyrinthine lair—with *Arachne,* the mythic spinner of webs, the spider. An early editor, Edmond Malone, noted that different editions may well have caused the confusion, or, as Malone seemed to prefer, perhaps the poet's memory induced it, if the fusion of two distinctly antithetical names arose "in his imagination." One wonders if Whitman's poem does not share the reverberating echo.

One ultimate adventure seems often too much for us: the force of cultural differences and desires—are they so violent that we dare not look beyond our own isolation in the world? One can only suppose that we humans need almost to think beyond ourselves, to survive, which would imply dreaming while wide awake.

II

The Mind Imagining

Although it is a precise mathematical method, topology pictures an extravagant playground for the mind, and it reaches out to many different formal arrangements, all of which help to determine the range of human behavior, as well as *abiotic* nonliving nature. From this wealth of transformations a reasonable conjecture emerges: the idea that today the world needs to value and understand the positive purpose of fiction and the imaginary—not simply more countable things and populations, more messages sent back and forth, more wealth for some, less for others. Counting in vast ramifications is the basis of modern science, and yet mathematics, however much its abstractions develop, cannot stand alone as a source of values.

Life is not just a branching accumulation of data to be analyzed; it is a body of interactions to be lived and then imagined as serving a purpose. Already in speaking thus we use complex words calling for interpretation, especially as metalanguage—words used to describe other words. The human species sometimes admits to craving a long forgotten resource—an intuition of our limits—and only within that inherently formal boundary is there a bounty of mind that folds and unfolds many kinds of intelligence, whether mobilized in the humanities or the natural sciences. Ordinary language can be used to reflect the subtle quarrels over the nature of these many approaches to the workings of mind, and in the present case we are merely suggesting that the rise of aesthetics in the Enlightenment period marks a new interest in what imagination entails. Central to this question is the problem of relating vision to form.

The historian of ideas alerts us that aesthetics, from the seventeenth century onwards, depends on the psychological notion that mind builds thought out of primitive perceptions and sensations, qualities of things both primary and secondary, and British philosophy is imbued with this loosely combinatorial theory formed from various modes of "association." Mind seems to grow from a life of connecting meanings to multiple notice boards, all flashing their messages for later use. To combine separate items in the logic of position, as Leibniz hoped, would build the array of its powers as an associationist machine.

There was nothing simple about such philosophic developments. Imagination is a disturbingly complex power, at least in the perspective of today's neuroscience, where researchers find links between it and memory. This was true when Shakespeare used the word in *A Midsummer Night's Dream* and no less true in 1934 when I. A. Richards lectured on the subject and published his pioneering book, *Coleridge on the Imagination,* leaving the psychological questions open for further critique, unlike poetic treatments such as Mark Akenside's genteel if enthusiastic *Pleasures of the Imagination,* published exactly 100 years earlier. Richards always knew how to avoid false critical enthusiasm, and he explored central questions about what constitutes imaginative thought or action, but to this day no precise or testable definition will indicate how the mind acquires and governs this power to reach beyond common knowledge. Terms like Imagination seem designed to obstruct rational inquiry, even though humans of advanced intelligence can sometimes recognize its relevance when they see its results. Of course, knowing how to acquire imagination would be instantly to destroy its possibility. Imagination goes beyond its own definition, and certainly beyond the material opinions and grasp of the way things are commonly taken to work. A much lauded capacity, imagining joins in a dance with fantasizing and even hallucinating, and yet when creative spirits call upon imagination, they bring to that call a certain discipline, taming the dream—no matter what may be the field of creative endeavor—art, science, or any other sphere. At issue, most often, is an instinctive Einsteinian or Flaubertian rejection of unexamined, banal, "accepted ideas," such that freely to define imagination, we are forced to look beyond the standard notions of such common paradigms and visions.

One underlying idea of this essay is consequently that creativity depends upon imagined links *between seemingly disjunctive fields of thought*—whether among words and discourse, images and narratives, numbers and their progeny, tastes and their refinements, or anywhere within the wide range of possible human experience. In my view these linkages are virtually all topological, since every serious encounter with scale and balance will require most humans to think beyond the narrow confines of those little worlds we know "for ourselves." To imagine remote possibilities is itself an exercise of the imagination; the probable possible, as Aristotle's *Poetics* suggested, is an imagined leap, akin to metaphor: something on the order of the machines envisioned by Leonardo da Vinci in his *Notebooks.* On the other hand, when and if we abandon local human knowledge entirely, we may fall into an error, by soaring too far "beyond the realm of possibility." Personal, limited, self-critical knowledge must always be given a strong role in the vision of the possible. Looking in two directions at once, into ourselves and out beyond ourselves, however ambiguously, we then must become the masters of imagination as it reveals the basic truths of our physical and ultimately spiritual existence. Vision in this extended metaphorical sense has a long history involving dreams, fantasies, and legendary lands. Nonetheless, I believe that the present ecological condition of planet Earth calls for a combination of scientific research, even though, ironically, it is as if we can measure but not see shape in the planet, though taking its measure. Significant new conditions of life continue to emerge daily, and global modifications of communication change the very idea of any settling local knowledge. The Big Picture is suddenly too small to make sense, which means that new regulations of power and more sensitive attitudes toward power in general are required by the species, and these powers are not laws, but changed patterns of vision.

The imagination seems to be a force of mind's creative power, which harnesses fantasy (as the ancient Greeks called it) and what I should rather call a *free play of mind,* something on the order of a controlled dream. There is indeed a long history of the attempt to understand such capacities. Northrop Frye rightly spoke of what he called "the educated imagination," following the lead of the Romantics. On a historical basis James Engell's book, *The Creative Imagination: Enlightenment to Romanticism,* chronicled the origins of modern wonderment, in Britain and in Germany for the most part, displaying a European belief in the role of creative vision animating lower to higher reaches of metaphysical and poetic ambition. To understand these virtually indefinable powers, let us think impressionistically, because a host of ideas comes to mind: imagination is picturing, guessing, supposing, daydreaming or simply dreaming, prophesying, hallucinating, fantasizing, schematizing, ruminating, even reasoning itself. Things need not reach such a dangerous pitch as Francisco Goya's "The sleep of reason produces monsters." In one version of the artist's *Caprichos*

43, Goya wrote another epigraph, "Fantasy abandoned by reason produces impossible monsters: united with her, she is the mother of the arts and the origin of their marvels," and then, adjoined to reason, imagination acquires its strongly positive sense. When in *The Phaedrus* Plato had long before allied imagination and fantasy with a "divine madness," he was specifically advancing a creative power and he was not denouncing the gift, despite suspicions of its wayward danger to any hierarchical, authoritarian vision of the state. However imagination expresses its powers, it always includes an unusual capacity to make connections between dissimilar things, drawing these together in ways that often violate common sense, in ways we recognize as artistically free. Indeed the freedom to imagine different states of things was exactly what both fascinated and disturbed the author of *The Republic,* especially in its tenth Book. To this day authoritarian governments have either been wary of poets or have entirely suppressed their professional tendency to think independently within the bounds of civilized society.

Personal freedom that implies controlled creative powers may be historically aligned with the eighteenth century beginnings of *The Aesthetic,* announced by Alexander Baumgarten's 1750 treatise of that title, but the connection belonged to an earlier development in Germany, later elaborated by Friedrich Schiller in his essay, *The Aesthetic Education of Man in a Series of Letters* (1794). A Romantic view of nature, especially a picturesque or sublime natural landscape, could be so intensely idealized that it generated a new pathos, raising a cult of "creative" imagination to a new place in art and life, and also oddly Hellenizing philosophy. The historian E. M. Butler called this a "tyranny of Greece" over German idealist thought, and we can only suppose that it accompanied a resistance to rising industrialism and the inevitable concentrations of growing cities. Ideas of nature and imagination were separated from the bourgeois charms, busyness, and grandeur of city life, a reflex epitomized more recently, as I have said, in the *volkisch* poetic philosophizing of Martin Heidegger. His quite special poetic diction is admittedly a thrill to one's archaic thoughts, but today the professional environmentalist does not care much, as neutral scientist, for nostalgic Arcadian legends nor pious communitarian myths of a refuge in Nature, and yet however much "the ecological imagination" has already been revived in current literary jargons, ecology does require a visionary extension of plain common sense, and the key word remains "visionary."

We cannot overestimate the influence of Romantic thought—think of a poet like Wordsworth—as stimulus to a belief that Nature's grandeurs and picturesque domestications call us to slow down the progress of heavy industry, which in turn would turn the world into the *anthropocene,* as thinkers today might dub our interference with natural process. As for the arts, it still pays to sing to the Moon. Although scientists continue an abstracting search for conceptual basics, along with instrumental techniques for measuring the physics of natural process, they gain inspiration from the great Soviet geochemist, Vladimir Vernadsky, when he translates the ecologist's biosphere into its imaginative form, the mindsphere, or, the *noösphere.* Vernadsky was well aware that humans have always throughout known history interfered with Nature, even if that only meant throwing stones at other animal species.

Here the truth of nature's process and reality, to adopt Whitehead's phrase, virtually demands openness to imaginative theoretical construction, since human invention changes natural conditions through artificial means. As the biologist Lynn Margulis maintained in several publications, Vernadsky saw that we must look beyond our own engineering and beyond our useful devices of external biotic controls—beyond our chemical fertilizers, as it were—in order to penetrate the deeper structures of the biotic world we inhabit, a world inhabiting us, the world of the extremely large and the extremely small. These topological extremes, as imagined, in terms of a human destiny on Earth live only inside us, that is, in our minds describing and our bodies experiencing them. Germs and living cells of all sorts of course quite literally do live in us, in our bodies, and hence they may be called our "inhabitants." The chief requirement of insight and wisdom in this ecological domain will be for most citizens to rethink, or begin to think, about what thinking actually is. Typically the recent worldwide scale of climate change puts the same pressure on imagination; it requires that we completely rethink our intellectual foundations, rejecting superficial short-term solutions, widening our picture of both time and space in human affairs, thinking beyond our own narrow personal interests. One can hardly plan this process of reaching beyond our own egos, yet somehow it must happen.

Among numerous senses Imagination now includes a sense that while the arts must inevitably call for creative genius, a similar imaginative gift is also to be found in abstract problem solving, as in developing sciences, or for interpretive linguistic purposes. Here the "poetic" model provided by Vico suggests a direction, since beyond the practical world yet another concern must extend our inquiry; it might be called "imaginative perceptions" or, as the poet Wordsworth named it, "intimations" of things that are out of sight. We know that the great poet's sister, Dorothy Wordsworth, trained him to see *into* the landscape that lay all before him, and this training of eye and mind required spiritual exercise. In the course of developing human intelligence, from earliest infant days into full maturity, even not very quick-witted humans make use of various mental capacities, including perception, orienting, reasoning, recalling, calculating, anticipating, estimating, and the like, while these and other cognitive capacities are attended by an array of rather more obscure mental events, all of which involve the realm of feelings, the full ensemble demonstrating a very wide range of emotive experience.

Psychology and psychoanalysis contribute to our understanding of the role of the fantastic in creative activities. Sometimes the more material functions of mind, such as measuring and calculating, are accompanied by mental events lying somewhere between thought and feeling, and one of these is what we call imagination. Consider a simple case: old-time carpenters, who could build a house or barn from nothing more than a rough pencil sketch drawn on a piece of paper, were in some sense imagining the actual blueprint a modern builder would be inclined to use. These old-timers had a definite "feel" for the structure they were about to create, and this feel has something in common with hand-eye coordination. In their schematic picture of the house, they could mentally hold actual, relevant measurements of wood beams, and, curiously, they were theorizing when they did so. The house would rise from the ground as a creative combination of vision and material fact. Like an awareness of one's own body, feeling its space and time and motion, a creative sense of structure is very much connected to the imagery that expresses contours of the real world, the world where perhaps the chief human endowment is the willingness to interfere with natural process. Henri Poincaré, whose dates (1854–1912) span two great periods of scientific accomplishment, was not without concrete material experience; when he began his distinguished career, he was asked to inspect conditions of coal mining, with its dangerous tunneling underground and its threat of underground fires. He later wrote that we orient our abstract thoughts by starting from the body image, and it is my belief that his experience below ground sharpened his sense of the necessary feel and his powers of imagining the shapes of space. The numbers would follow; the shaping of the space came first, and in this

Poincaré cannot fail to remind us of Gaston Bachelard and his inspired phenomenological books on the elements, such as *The Poetics of Space, The Psychoanalysis of Fire,* and *Water and Reverie.* Bachelard was a scientist by profession, but his books reveal an underlying link between creative thought and the role of displacements in a psychoanalytic sense. We might well analyze this crossover by considering the dream. Freud recalled in his 1900 classic, *The Interpretation of Dreams,* that dreams were defined by the ancient geographer and diviner, Artemidorus Daldianus, as "the thoughts of a man who is asleep"; here imagination seems to remain vague but vivid, profoundly ambiguous in its meanings and hence its interpretation. In the dream, as in deep mythology, there is a constant seesaw between the intense image and the turbulent logic of psychodynamic displacements. The dream links to primordial beliefs, such as the Myth of the Eternal Return, as already suggested. The dream belongs close to imagination and fantasy in the deeper analysis of mind, and to this extent our psychology must deal with aspects of the Freudian Unconscious, to use a terminology still powerful, if not fashionable at the present day. More than one hundred years ago, Freud's 1900 volume, *The Interpretation of Dreams,* showed that dreams operate with perpetual metamorphoses through displacement and condensation of imagery drawn from the residue of everyday experience, but when he wrote on the creative literary arts, he entitled his modest essay "On the Poet and Day-dreaming." Daydream is every bit as complex as those thoughts of a person dreaming in sleep. Early psychoanalysts found a creative freedom specifically at moments of awakening and of drifting off to sleep, where the Unconscious proper is not yet abandoned or fully in place. Wherever artistic shaping occurs, it is subject to psychoanalytic displacements, negations and condensations, as Freud showed. The artwork has its *secret* craft, as David Hume once wrote, describing tragic drama. At either deep or surface levels, the dream demonstrates how the mind may combine and recombine psychic materials with a virtually abandoned freedom, and yet when identified with poetry (the ancient art of *poesis* and the thing made, the *poemata*), creative art needs or at least should require specific and serious trained artistic skills. Only thus, only by cultivating modes of artistic order and expressive technique, will creative drives sustain imaginative play as a function of mind lying beyond quantity and beyond the reach of any positivistic experimental psychology. On this basis we might say that both art and science turn the dream into an available daydream, so that the latent energy of the unconscious becomes an openly available energy. Imagination requires a combinatorial sense of disparate realities, especially those in principle overlapping in their perceptions.

If perception as well as cognition is influenced by underlying imaginative "seeing," we may return once more to our main proposition, that a shape-conscious topology connects directly with imagining invariant shapes, and thus also with the obvious shaping of the work of art, which such artistic work always intends. Imagination is not a forest of image-making wilderness. We can go even further, if we think how shape can be represented by words, to suggest that topology is also the main bridge joining mathematics and the literary use of ordinary words, while both "languages," as Galileo long ago suggested, deploy a considerable range of ordering styles.

From such collocations there emerges a major theme, namely that we profit by associating the powers of science and art with each other. Suppose we extend somewhat Wittgenstein's idea of language games and their attendant problem of rule-following: we arrive at an extended version of Galileo's famous statement (in *Il Saggiatore*) that mathematics, which mostly for Galileo meant Euclidean geometry, is "the language of science." In a personal communication Bradley Bassler has described the union of art and science in this way: "Both art and science are primordially dependent on an underlying shaping power that comes from deep sources in our affective and cognitive orientation in the world. The expression of these deep orienting roots in the arts and sciences is a kind of high-level mental recycling of our most fundamental resources, used daily in our ongoing human experience, but here elevated to the level of products both themselves fashioned as Freud showed, out of the detritus of previous consciousness and yet also functioning as templates for future communal enterprise, a kind of grand recycling on the level of social exchange." Topology had not yet been invented in the seventeenth century, but its rudiments may be discerned in various expressions of mindful judgment. While for most people the idea of imagination applies chiefly to the arts, philosophy, mathematics, the natural sciences, and most decisions that guide ordinary living involve such a search for available imaginative energy, ordered by invariant shapes of use, and they will supply many reasons for extending the necessarily ambiguous range of our central terms. Despite any difficulties of definition, we do recognize creative imagining and we readily speak of its hypothetical character in various ways; we call it creative, ingenious, inventive, gifted, and so on, but with these qualities we always must include the major quality—the play of formal invention.

Our language games here all suggest that the power of felt imaging is some sort of advance upon merely referring to things, or naming events, or even perhaps of measuring the quantities. Hence, certain operations of algebra lead far beyond the simple situations from which they start; at times the mystery lies in the darkness of apparently crude initial conditions. A human's whole life can be ruined by being born too tall, or born on the wrong side of the hill. Geography or accidentally meeting strangers at the wrong crossroads, or boarding a different ship, can utterly alter a destiny.

Great literary works have been composed on just such grounds of myth and logic. Nor is the history of abstract thought immune from such chances. Some of the apparently simplest mathematical properties turn out to have similarly complex consequences, when developed; an example would be one of the most interesting intersections discovered in all thought, Richard Dedekind's (1831–1916) conceptual cut between any integer and all the numbers smaller and all the numbers larger than the originally chosen number. Arithmetic altogether seems to depend upon this *Dedekind Cut,* and when we ask how the vast construction of arithmetic holds together, we must believe the force of the formal principle. Sometimes the most complex physical events arise from the simplest initial (butterfly effect) conditions. This is nowhere more obviously the case than in the field of biology, with all its ramifications as revealed by our microscopic instruments, nor is it easy to avoid the history of organicism in Romantic literary forms which aspire to a symbolic imitation of natural organic processes. The Romantics saw that if things are alive, they must be participants in a biosphere, which in turn means that the Romanticism is deeply involved in a difference between rational consistency (an Enlightenment goal) and an actual environmental coherence. Eugene Wigner, a Nobel laureate for his work in quantum physics, asked in a celebrated paper how mathematics could prove so "unreasonably" useful in the "abiotic" physical sciences, and on persistent reflection Wigner found no ready answer, although it does appear that the relations of material forces, masses, particles, motions, and other events alone knowable as fact—that is, exactly demonstrable—are only thereby known to the scientist in mathematical terms.

Yet Wigner's single word, *abiotic,* referring to the nonliving materials of physics, suggests one answer to his own question: buried just beneath the surface, physics assumes a troubling qualitative condition, namely that *nonliving* matter lacks anything resembling consciousness (such as we intuit our own states of mind to possess in their ever changing flux). Quality belongs with consciousness, and this in turn means something humans possess. Sir James Jeans once remarked that if God is anything, he must be a mathematician. Certainly cosmology would support such a guess. Supposing this metaphor to have weight, mathematics itself has such variety that no simple dichotomy will do the field justice, yet the invention of topology does support a split between two different modes of consciousness: the discipline either counts things, or imagines them, or both together, as my argument contends. Of course, mathematicians like Georg Cantor, when picturing the formal properties of infinity, do indeed sound like a reduction of kabalistic mysticism or a Kantian "mathematical sublime," as described in the *Third Critique.* On the other hand, imagination may display sudden theoretical vision, at which moment a Cantorian freedom in Wallace Stevens' late-life poem, "The Planet on the Table," comes sharply into focus, at once both real and abstract.

Albert Einstein, when describing his own thought processes, once revealed that only rarely did they consist of words, but instead were ordered "in some kind of survey, in a way visually." One might suppose that natural languages, while expressively powerful, fail to reach a high enough degree of precision for use in the physical sciences. Mathematics seems to be the only semiotic system capable of providing exactitude in describing nature, although the violin-playing Einstein was careful to add the important qualifier, "in a way." Like other scientists and artists, he never ceased to wonder at our indescribable powers of discovering the *qualia* hidden by the Natural order.

Nonetheless, everywhere we look into the play of high mental achievement, we find that the role of imagination calls for a better understanding than has been available, even among artists, not to mention scientists. A Californian mathematician friend (if such thinkers may truly be located geographically), who works in the field called "Analysis," once said to me that whenever the infinitely extensible field of mathematics expands to higher levels, "It is not about anything at all." Of course, it refers to its own expansion. If so, he then observed, rarefied mathematical models may occasionally be drawn back down from the higher levels of abstract implication, where purity implies a possible useful application, but this descent to the real world occurs only when an alert scientist notices a potential use for the model.

In a chapter on basic principles, Barry Mazur's book, *Imagining Numbers,* tells of the complementary relation of discovery and invention in math, where the idealist philosopher Fichte speaks of a Platonist freedom and imagination in *discovering* the hidden truths of eternal mathematical order existing "out there," while on an opposite plane of intuition the mathematician may experience the thrill of *inventing* the forms of this order. Although discovery and invention are close twins, they share in a higher imaginative reach, no doubt always requiring abstraction, as Alfred North Whitehead wrote in a chapter of *Science and the Modern World.* To the extent that science and technology are currently mesmerized by questions of scale—the very small and the very large, with their small and large corresponding numbers and magnitudes—there is bound to be an imaginative use of abstract mathematical models. Higher mathematics flies high above adding and subtracting, not only in more complex forms of mathematical proof, but also when those rarefied conceptions and proofs are applied to complex physical systems, such as we find in quantum theory or in cosmology, and yet we are always adding, subtracting, or multiplying. The vastness of Pascalian space enforces a betting table for imaginative thinking, and yet the scale of the cosmos is not where our highest powers of this kind should be mobilized. As we shall see, the cosmic scale produces analytic problems of virtually irrelevant scientific activity, despite the perpetual public relations campaigns celebrating Nobel Prize winners and currently glamorous fields of research, which today would include neurobiology as the promise of explaining consciousness, and so on. By asking wherein lies the invariant aspect of change, I am essentially asking what is relevant to the life process.

The doors of imagination open upon many kinds of thinking, not only nowadays, but clearly that was always so. Ancient Greek philosophy noticed how human thought enjoys a kind of mental athleticism when exercised on a high level of contact with the natural world, and the pre-Socratics understood that concepts of Nature had necessarily much to do with her observers and interpreters, namely humans thinking. The earliest Western theories of mind assume a partnership of thought and feeling, and we need to reserve judgment, if we think that all invention and discovery is the child of pure reason, since at its most inventive, human thought always strives beyond common sense boundaries, while early science gradually reveals that common sense may be seductively misleading.

Meanwhile, wonder remains almost always an active participant in the ancient philosophy of mind, as for example in Plato's dialogue, the *Phaedrus,* which describes the four types of "divine madness" or mania—(1)

prophetic, (2) mystery-ritualistic, (3) poetic, and (4) erotic—expressive powers each consisting of an irrational daemonic energy and vision inspired by the four divine Powers, Apollo, Dionysus, the Muses, and Aphrodite. We are not shocked that our ideas of science play little part in the Platonic (and other similar, let's say, Gnostic) accounts. But science is thrusting outward from the ancient mysteries of mind and sensibility, and in many forms of poetry the mind stretches forth, questioning the literal heaviness of common sense pragmatism. The poetic and broadly artistic approach to representing reality inspires humans to develop language, and language with all its rich syntax and semantics, whether verbal or mathematical, becomes the critical human skill for all attempts to express and eventually solve the riddles of existence.

While in the modern world we identify progress with science and technology, ancient astronomy and Euclidean geometry took their first critical steps in the ancient world, millennia before the sixteenth century. In many essays Gerald Holton and other historians of science may best be understood as students of a special branch of imaginative activity, for there can be no doubt that the capacity to discover the logic inhering in the Pythagorean Theorem would be felt as a kind of "seeing what was not there before," a mental leap beyond the materially visual aspect of the triangle. Yet the very power of envisioning forms underlying reality carries logic ever higher, as subtle conceptual derivations come to mind. There are obscure shadows in these varieties of imagination, already envisaged by Aristotle in the fourth century BCE, when he associates knowledge with more than one imaginative domain, although one finds in his works a strong leaning toward the study of biological process. Even in his *Poetics* he treats the poem as a kind of probabilistic activation of meaning, and as we shall see, he particularly praised the rare gift of metaphor, which he said cannot be learned, no doubt because this gift assumes a poet's contact with life and biological stimuli on the one hand, and on the other hand with a peculiar hyperactivity arising out of such contact. In metaphor languages adopt a new identity, if only for a moment, only to vanish, as system, in the next moment. Every strong metaphor questions the conventional stability of some worldview, by noting that it is we humans who decide what shall be our standard picture of a universe far beyond the range of ordinary comprehensions. Once again we find ourselves using that telltale term, beyond, and once more we ask if this "beyond" does not imply a special boundary of sight and insight, the horizon.

In critical discourse the idea of creative discovery takes a great leap forward with the later rhetorician, Longinus, whose treatise *On the Sublime* (undated, but perhaps written in the first century of the Christian era) directly links cosmic grandeurs with the greatest imaginative power, as if the highest poetry were a mode of cosmology. While Aristotle connected Sophoclean tragedy with mimetic or "realistic" renderings of probable or possible human experience, Longinus much later soared higher, quoting "Let there be light!" from the Bible, as an example of sublime gnomic utterance and cosmic vision. With him we seem to have suddenly anticipated the modern Romantic period; to this end Richards quoted Longinus (Ch.35) in *Coleridge on Imagination*:

> Wherefore, [says Longinus] not even the whole universe can suffice the reaches of man's thought and contemplation, but oftentimes his imagination oversteps the bounds of space, so that if we survey our life on every side, how greatness and beauty and eminence have everywhere the prerogative, we shall straightway perceive the end for which we were created. (1956, p. 24)

This excerpt displays an unbounded cosmological stretching of imaginative scale, which will always occur when the object of wonder is the whole universe itself, a sublime object so great, as Immanuel Kant would claim centuries later in his *Critique of Judgment,* that sublime magnitudes must absolutely overstrain the mind's powers of analysis and response. Imagination, at least in relation to the sublime, has the widest possible scale of mental challenge. As Gerald Holton suggests, when considering Einstein's overwhelming task of relativity theory, this modern version of the Longinean imagination that "oversteps the bounds of space" is rediscovered as a special kind of poetry, when early stages of theory are expressed as the scientist's "thought experiments" (1996, p. 89) Einstein is generally accorded a supreme gift of specifically topological imagination, but thinking of Holton's comments, we must also note that the classic epic adventures in literature, such as *The Odyssey* or *The Divine Comedy* or *Paradise Lost,* are similarly extended epical thought experiments.

To mention these great poetic works, composed during such an extended historical span, is to accept that our topic of imagination has an equally rich and extended history, and for my purposes it is only a slice of modern times that can occupy this account.

Coleridge, Schelling, and the Two Levels

Galilean demonstrations and mathematics changed the grounds of modern thought, and by the end of the Enlightenment Period modern science was having an immense influence on all the currents of thought; the world was essentially "mechanized" after Galileo and Newton, and the nature of transcendental vision was everywhere being reexamined, especially among German philosophers. With the rise of German Idealism in the eighteenth century, it became clear that humans might wish for an escape from what Jonathan Swift memorably called "the mechanical operation of the spirit," with the result that imagination, while associated with or deeply rooted in poetic power, was also in every respect linked to the question: how does reason connect with knowledge and subjective consciousness? Georg Lichtenberg (1742–1799), a scientifically active physicist and brilliant polymath, could produce aphorisms like this: "The most heated defenders of science, who cannot endure the slightest sneer at it, are commonly those who have not made very much progress in it and are secretly aware of this defect," while ironically adding in another of his famed *Aphorisms,* "The pleasures of the imagination are as it were only drawings and models which are played with by poor people who cannot afford the real thing." Who's there, one nevertheless asks, once again standing on the darkened battlements of Elsinore, and one asks who is imagining what? Laced with irony, the debate surrounding the scientific hegemony—the Galilean pursuit of hard cold fact—was to continue until the present day and will no doubt continue for a long time to come.

In the British context, as the historian Basil Willey showed in more than one book, by the early nineteenth century the Continental idealist tradition had reached a critical stage of development, when Coleridge produced a famously suggestive but incomplete account of the full range of imaginative powers, as shown by Basil Willey's *Samuel Taylor Coleridge.* A period of European literature that included Goethe's *Faust,* especially *Part Two,* was not likely to neglect poetic cosmology, while for Coleridge imagination and perception were closely allied functions of transcendental thought—a belief acquired in part from sharing the study of natural landscapes along with William and Dorothy Wordsworth, the poet's percipient sister and author of an epochal, if all too brief, journal. Interestingly, they lived in what is called "the Lake Country."

When the mind actively encounters the objective outer world, its landscape, there is an inevitable tendency for subject and object, feeling and thought, to collapse into each other, a coalescence defying the Cartesian dichotomy of thought versus material extension, the *res cogitans* versus the *res extensa.* As will appear, following our central dictum that proper decisions require proper incisions, to break the stranglehold of Cartesian dualism, we shall need to examine the conceptual edges that define research into meaning. However the imagination engages altered vision, we can see that it verges on a sublime Longinean freedom of reference, since it obscures any neat conceptual wall between one thought and another, one object from another, radically destabilizing each extended area of existence, locating it beyond any familiar *either / or* demilitarized zone. In more specific terms which I shall soon examine, uncertainty of boundary leads to an essentially metaphoric, interactive enrichment of meaning, when imagined relationships are seen as part of some continuously creating process, a *natura naturans.*

Nature is oddly (we think) a social invention guided by the use of certain language games, particularly those of bent and stretched meanings, and these expressive shapes need to be revived, because many of the children of the new technology, we might call them the algorithmic generation, keen to be networking, will finally join forces in desperate searching for lost community, as utopian philosophers have consistently warned. Such would be the confusion of quantity and quality, a course of history Vico named the "barbarism of reflection." This unhappily vicious, sophisticated fate the Romantics warned against, as witness William Wordsworth's famous lines:

> The world is too much with us,
> Getting and spending, we lay waste our days.

There is certainly an affinity between such Romantic views, both recreative and procreative, and the fact that the Romantics in Europe were much concerned with the arts, and they were not yet engulfed by an all-powerful establishment of the physical sciences, although some, like Percy and Mary Shelley, not to mention Goethe, were often keen students of science and much admired its progress, an advance calling for a new metaphysics. Literature for them often indeed had close ties to the new sciences.

One expression of this rich literary understanding stands out: To understand the play of imagination abstractly—as opposed to recognizing it intuitively in a work of art, let's say—besides noticing that it virtually

preferred by us to an uncivilized" (1956a, p. 58). These phrases themselves evoke all the sharpest moral dilemmas of the present century and quite evidently indicate that to imagine is to think, which is partly what the Cartesian *cogito* implied, along with a measure of emergent inspiration, when Secondary Imagination "dissolves, diffuses, dissipates, in order to create." Its force is "is essentially vital, even as all objects (as objects) are essentially fixed and dead." Given that view, SCT calls imagination an *esemplastic,* infolding power to unify elements thus dissolved away from their fragmentation. We continue to notice that when measurement of magnitudes is not allowed, as Leonhard Euler insisted, the rigorous thought of the topologist must focus on processes of potential flow and local adjacency, and then also of homology *(homeomorphism)* among shapes. This latter approach to plastic space strongly resembles the late attempts of Whitehead in his strange book, *Process and Reality,* where even triangles have feelings and things are actually events.

No doubt the most illuminating account of all the above mysteries is David Bohm's article on "The Range of Imagination," which later appeared in his collection of essays, entitled *On Creativity.* Bohm begins his account with a careful treatment of Coleridge as interpreted by Owen Barfield, and he goes on to show how the modern theory of light and also the method of using ratios will enrich the original Schellingian definition. This widened perspective allows Bohm to describe various ways in which science can arrive at a "higher" mode, which rises above fancy and its mechanisms of association, to permit the *esemplastic* unifying vision of things. Wider and deeper questions raised by the Imagination essay were developed in the remarkable 1980 volume, *Wholeness and the Implicate Order,* where Bohm elaborated his ideas concerning "the unity of the universe," as it is commonly called by scientists, here an "implicate order" specifically linked to Bohm's own expert field, quantum theory. As topology preserves resemblance in change, when an object is stretched or twisted or similarly deformed without tearing, so the Coleridgean esemplastic power, by looking beyond what can be counted, abandons measurement in the interests of identifying shapes, and these are seen as states of potential metamorphosis. Clearly, if the universe is a One, a unity, as Denis Sciama's *Unity of the Universe* had dubbed its wholeness, then the universe has or simply is a shape.

At least within the Romantic orbit of organicist creativity, the imagination on its higher levels of thought and feeling attains a topological phase-change, molding the materials thus reshaped into a more flexible union of previous fixities. In this way imagination raises materials into a relation with ideal forms, as the poet raises the plain sense of things into a complex idea of their essence.

A Local Habitation and a Name

If we accept the idea that cultural changes reflect an underlying creative human power, we can say that imagination depicts causation as a metamorphic process—carrying our thoughts beyond mechanism, and hence beyond the Newtonian worldview—while fancy deals in the discernible reasons for connecting already known things or facts, the Coleridgean "fixities." Few authors have ever been such violent opponents of Newtonian celestial mechanics as William Blake, but as historians of science like Peter Galison and literary critics like the philosophically oriented Thomas McFarland have written, Coleridge and other Romantics discovered that visionary freedom included the mystery of both the object and also a subjective, if abstract notion of object as function of consciousness. Like an alchemical process, the Romantic object does seem to project a Symbolistic aura, like a jewel or an ocean surface in Baudelaire. That is what STC meant by *esemplastic* fluency. The making of imagined objects, whether in works of art or strange scientific theories, is perhaps the best study, if one is to grasp the inherent import of topology so far described. For Poetry is a making, as the term *poesis* means in the ancient Greek language.

Meanwhile, on the lower level of Fancy one discovers what perhaps are the simplest reasons and causalities that explain machines of any kind, even including the Newtonian cosmos—we recall that, regarding his system of cosmic time and space, Newton could speak of "God's sensorium." Poetic imagination may rise above Fancy and therefore the poet takes a humbler stance, as when we find mere humanity in King Lear's enraged outcry to his daughter: "Reason not the need!" That feeling catches the rock bottom sense of "reason," to be sure. Meanwhile Lucretius' poem *On the Nature of Things* had taught Epicurean poets to accept relentless, creative, swerving turbulence, while Ovid's *Metamorphoses* taught poets like Shakespeare to envision in human life a process of universal shape-shifting, as if the divinity that "shapes our ends" were the Native American Trickster. One might wonder whether creativity is not a partial grammar, an immediate consequence of ordinary language trained to conventional styles. As Cassirer and other mathematically trained philosophers have said, culture seems to develop from early periods of *mythos* to later periods of the *logos,* but without a bending freedom we cannot ascribe creativity to rigorous logic, as appears in the drama.

Our theme of imaginative freedom might be seen to lead almost anywhere, even though the Enlightenment studied its conditions with great earnestness. A famed composer later remarked that even the apparent rationality of familiar social arrangements can severely strain the relation of causal process and experiential truth, so that we get "effects without causes." The composer was the anti-Semitic Richard Wagner, of whom we can believe almost anything, but even so, although Meyerbeer has his spectacular lack of form, bizarre operatic special effects do suggest or imply causality. The idea of cause is famous for its grip on the philosophic mind, as a problem of deep significance. Nonetheless, lighting a fire does suggest an actual proximal cause, from which the fire itself is not exactly "free." Despite this commonsense view of process in the familiar macroscopic world we inhabit, philosophers following Hume and scientists in the quantum field of physics have argued that cause leads to effect only because it is a crude statement of an erroneous impression got from experimental correlations.

In the late sixteenth century Sir Philip Sidney's defense of poetry asserts a fundamental artistic principle: the poet in his divine madness "nothing affirmeth"—like all discoverers of new forms, the poet shapes fresh fictions of factual representation, which is one of the first consequences of his "freely ranging within the zodiac of his own wit." Such a refusal to "affirm" is the hallmark of the greatest English Renaissance poetry, and the declaration of mental freedom applies equally well to the great scientists; it applies to Galileo, and to his admiring follower so much later, when Einstein rejected the standard nineteenth-century belief in the existence of the Ether. He said that we cannot affirm the existence of the Ether waves, because they simply do not exist. In a sense he was affirming a nothing; his negation took a positive step.

Supposing that Thomas Kuhn's *Structure of Scientific Revolutions* correctly shows—at least in part—how scientists building on the past can still break with current dogma, thus changing the paradigm, even more readily will we find this happening in the Arts. Scientists have more than once disagreed with Kuhn's idea of paradigms, but the Arts give us a model of imagination in its most mimetic or most expressive and hence most recognizably human form, and when we grasp the nature of this metamorphic vision, we will have seen how all the most important intuitions of fact as well as value are visions of orientation changing in time. Perhaps if we trained ourselves to feel the odd inevitability of the one-sided Möbius strip, we might learn more about the ways the mind combines all the thoughts it in fact does combine, we would "understand" the processing of Imagination—but outside a topological approach that is not possible. Nonetheless, there is every reason to believe that historically the cultural readiness of the West to assimilate the science of topology owes much to a Coleridgean esemplastic power of mind. Hints of this affinity appear throughout Christopher Marlowe's visually oriented tragic drama, *Doctor Faustus,* anchored on the famous rhetorical question about Helen of Troy, "Is this the face that launched a thousand ships?" Influential throughout Europe, this play later reinforced the Oswald Spengler label "Faustian" for extreme modern scientific daring, by making a hero of the studious scholar who is willing to enter the precincts of forbidden natural knowledge. The tragic tone returned in the twentieth century with Thomas Mann's last great novel, another *Doctor Faustus,* but now using a pervasive German musical motif to cover all of Marlowe's original dream of excess. Here the hero Adrian Leverkühn's excess is literally a breakdown of his physical and mental health, his failure to "stay in good shape," to use the common phrase. For complex personal and literary reasons, Mann throughout his career identified good health with the organic unity of life in general, and that belief he could surround with conjectures in his earlier novel, *The Magic Mountain.*

Imaginative literature, freely invented, praising discovery, seems often to presage similar science-based insights, as perhaps Dante's *Inferno* anticipates Newton's inverse square law of gravitational attraction, or *Paradise Lost* presages the 1915 General Theory of Relativity. Poems give humans a metamorphic schooling into the range of imaginative freedom, for poetic making implies myriad inventive possibilities, whenever imaginative wondering plunges our powers of reason into the sea of causes and empirical and emotional forces and drives. Literature, it needs saying to scientists, differs from music in that its language forces us to engage in thinking about what we loosely call "the real world." Contrasting the dream and the real, we find it intelligible enough to instance Shakespeare, whose later plays especially initiate the Romantic tradition of drama throughout Europe as much as with British literature. He understood and knew how to enact all the passions of mind and heart, and his seemingly calm Prospero in *The Tempest* knows his own turbulent ocean, his own inward tempest, his own "beating mind" that he wishes he might cajole nature into a calming season. There is almost no way to avoid the liberating influence of

Shakespeare, in all this change from medieval to early modern thought, an influence all the more powerful for many Europeans (even the French, who were not always happy about it) because the poet maintained a studied conservatism of ethical values and a shrewd political neutrality, knowing that Poetic authority lies elsewhere.

Regarding imagination we are fortunate in that, as English speakers, one dramatic exordium remains memorable and always worthy of close attention—the speech artfully delivered by the mythical prince of Athens, Duke Theseus, in the closing act of *A Midsummer Night's Dream.* The lines cut to the heart of imaginative freedom, when the Duke speaks his lengthy prologue to an archaic rustic play to be performed by Bottom the Weaver and his "mechanical" friends. Their performance will be of an unwitting comedy, and no less profound for all that. However, addressing the Court before the players begin, the Duke pronounces a fairly complete poetic theory:

> I never may believe
> These antique fables, nor these faery toys,
> Lovers and madmen have such seething brains,
> Such shaping fantasies, that apprehend
> More than cool reason comprehends.
> The lunatic, the lover, and the poet
> Are of imagination all compact.
> One sees more devils than vast hell can hold.
> That is the madman. The lover, all as frantic,
> Sees Helen's beauty in a brow of Egypt.
> The poet's eye, in a fine frenzy rolling,
> Doth glance from heaven to earth, from earth to heaven,
> And as imagination bodies forth
> The forms of things unknown, the poet's pen
> Turns them to shapes, and gives to airy nothing
> A local habitation and a name. (Act V, scene 1, lines 2–17)

Duke Theseus is exploring the nature of the lover's mania, the dream and its *dreamwork,* as Freud was to name the unconscious psychic process, but he is finally anchoring this dream to the arts of invention, most notably poetry itself.

The rhetoric of climax means a lot here; the key to this rich passage is the doubling phrase, "a local habitation and a name." The most important aspect of our intuitive knowledge of who we are is thus *to name where we are,* what place we inhabit, and when the Duke takes poetry to be the central source of personal authority in these imaginative domains, he locates the dream inside the dream, a shaping of vision that occurs in sleep, as in Artemidorus's ancient, comprehensive definition: "the thoughts of a man who is asleep." If the dream names all parts of this particular dramatic action, its emphasis on naming defines a true locality, a real dream of position on Earth and in society. Such is one necessity of true locus, of local habitation—the dream is home at heart, and thus dreaming, like love, is inherently always uncanny, *unheimlich,* and yet this is an estrangement from and to which we wake each day. Such interpretations show that longer works influence their own development in later repetitions of earlier expressions used in that longer form. There is a certain critical irony at work here. The Duke's prologue to an action, to the rustic play and its rude players, seems to draw a direct equivalence between their fantasy and his imagination, between dream and vision, but precisely because the passage was rhetorically canonical in English literature; Coleridge shaped his own definition in disagreement with the Duke's words, which spiral outward. As many have noted, the Duke's speech echoes manifestos like Sir Philip Sidney's claim that the poet is lifted up into nature-making, like some godlike demiurge in Plato's *Timaeus.*

The poet's fine frenzy power is the Demiurge, a creator spirit "freely ranging within the zodiac of his own wit." By itself and by its inspired melancholy solitude of vision this fine frenzy of a divine madness achieves a translated state of mind, its imaginative productions becoming in effect "another nature." Only thus can art dissolve, diffuse, and dissipate, in order to recreate an original creation, aligning science and art. Poetry has its own laws, however, suggesting a reason for Hermann Bondi's word, when he says that time is *mingled* mathematically with the three dimensions of space in Einstein's 1905 Special Relativity. That time may be a fourth dimension should then be regarded as a mathematically troped or mingling metaphor. Finally our idea must be that time both is and is not the fourth dimension, Prince Hamlet both is and is not crawling on the earth—a paradox or wrenching contradiction discovered by all who invent original conceptions and ways of life. Mind itself seems to be a coincidence of opposites, and metaphor is designed to entertain these conflicts. Vico held that what we have made, as cultural poets of any and all kinds, is what we truly know, and by the end of the nineteenth century we are left at the doorstep of Oscar Wilde's epochal essay on "The Decay of Lying," which argues ironically for the realism of fictions. With such Herodotean power to bend history goes Wilde's emphasis on the inevitable fiction, ambiguity, and relativities accompanying any humane judgment, while, by contrast, twentieth-century physics inaugurates a period of deep lying uncertainties in all areas, whether poetic or scientific.

In the field of the imaginative arts such uncertainties relate mainly to the use of metaphor, and the play of metaphor and other similar figurations of thought make imaginative combinations possible. Robotic algorithms have no such imagination. Alberto Pérez-Gómez in his *Built upon Love* and elsewhere has questioned whether a technology divorced from the realities of human desire will not enfeeble our ethical sensibility to the point of undermining the very aims of techological advance. On this view, if taken to be an absolute ontological good, technology becomes a dark threat, and with the semiotician Umberto Eco we must defend the scandalous mind-play of metaphor, which breathes paradox and discontinuity into our ideas. These disparities are the islands of thought, and they join, by contrasts, with the main bodies of that thought.

III

Disparities in Metaphor

Consider, for a moment, the current linguistic conditions for the use of metaphor. We nowadays live, depending on a questionable if not sardonic viewpoint, in a catastrophically new belief that information has magic, its source in physical observation of fact constituting and confirming a faith. In principle authority of any sort implies an augmentation of order, so that in our new digital age, where there is quite simply too much data to be surveyed, one aspect of what has been called *infoglut* is that we can only pretend to understand it in philosophic terms. Pragmatically, we do not easily handle such large numbers of "facts" and knowable data. Information theorists may comfort the general thinker by invoking the logic of Bayesian Surprise to determine which information is truly significant, but so far we are losing perspective in this self-replicating maze of "more is not enough." We already can see that "diaphoric" definitions of data attempt to treat fuzzy information that does not quite fit, as if some facts refused to accord with the continuous stream of already known standard factual arrays; but once again the glut—its imperial quantity—itself is the real point, because the data-gathering almost always looks for new ways to reinvent its own wheel, like a bureaucracy, and this compulsive hunt fails because it rejects the implications of ordinary language, which instead works by interconnecting endless unfamiliar ambiguities.

Under all sorts of literalizing conditions there would hardly seem any utility or truth in the awareness and use of a "bent" reality, and it follows that thinking based on metaphors could only be suspect, at best. An argument needs to be made for a programmatic sensibility, where ambiguities—all seven types, and more beyond—are accorded positive value. This positive quest would seem to need assessment and survey lying beyond countable quantities, something more like the "shape of things to come." Without claiming that we can exactly measure this need, indeed accepting that this is the real point of the exercise, we also need to discriminate the contours of such shaping. We need a countercurrent in our speech, which is exactly where metaphor and similar figurations play such an important role.

Humans in their use of language have long noted that a growing mass of specialized terms underlies troubles with meaning in general, and also that dictionaries tell only the superficial story of meaning. Metaphors, properly understood, alert us to this source of constricting literalism, and in 1936 Richards's *Philosophy of Rhetoric* emphasized this literalist modern problem for the reader. He worried that we would lose the feel of "The Interinanimation of Words," as if one could make an enthusiastic yet principled endorsement of the vastness of natural or so-called ordinary language, even when specialized. Underneath this vastness there is a built-in function of semantic tensions, a language of disparities.

Ordinary language works by hybridizing meanings, and this in turn suggests that a fully developed language amounts to, actually models, an environment of living organisms. Richards revolutionized the theory of metaphor through his terms "tenor" (loosely, the meaning) and "vehicle" (loosely, the metaphoric bending of common references and overtones of meaning), and while tenor and vehicle need a lot more scrutiny than they have always received, they convey the chief Ricardian principle, namely that language in use is a dynamic active process. It is not like tacking a family snapshot on a plaster wall, which would be a fanciful metonymy at best, although I am well aware of Walker Evans's famous anti-metaphorical—that is, metonymical—photographs in *Let Us Now Praise Famous Men* or his later subway photographs, images which raise the question: is photography not through exact representation forcibly destined to be a documentary device of metonymy? Evans assumes a stark subversive novelty in the photographic act of unmediated seeing, of looking without thought of meanings lying beyond. He assumes and uses that unquestioned contrast. Instead, with great precision and charm, Harry Berger's recent *Figures of a Changing World: Metaphor and the Emergence of Modern Culture* widens the scope of inquiry and raises another larger question: much earlier, was not the massive shift from medieval culture to modern mechanized culture effectively "the transformation of metonymies into metaphors?" (2015, pp. 33, 33–39) With Evans's great photographs, that slide into metaphor was often reversed, so that the sharecropper's boots would hover halfway between metaphor and metonymy. That would constitute a twentieth-century return to an earlier age of transition into a Shakespearean figural culture. The fact-like feel of metonymy begins to lose its allegorical force. Certainly, as a crossover into the early modern period, to follow Foucault's suggestion in *Les Mots et les Choses,* the Renaissance period is steeped in a cultural taste for metaphoric affinities—the perfect transitional mode at that time. Such a change amounts to a new way of imagining the order of things, obeying a collapsing hierarchy of broken Ptolemaic cosmology.

Proliferating metaphors and elective affinities open the doors into the New Science, but also into a new art and poetry. A cloud of interpenetrating metaphors fills the sky with questions about what we really do *see.* Instead, modernity insists that we do not always get what we see, since with metaphor extreme ambiguities result, as Richards's student, William Empson, a trained mathematician, demonstrated in *Seven Types of Ambiguity.*

Language at large seems radically extended for English in its Renaissance period (unlike French, for example). Literal meanings are disturbed but also enlivened by metaphors. While most traditional ideas of metaphor had stressed the analogy or likeness between different entities, the Ricardian view held that metaphor possessed a cutting sharpness of expression which derived from an antithetical troping of tenor and vehicle: "*There is disparity action to their resemblances.* When Hamlet uses the word crawling [to refer to human beings] its force comes not only from whatever resemblances to vermin it brings in, but at least equally from the differences that resist and control the influences of their resemblances" (Richards 1965b, p. 127). Troping, as we shall see soon in more detail, is a twisting of the sense of the words, so that their idea or meaning acquires a different cognitive shape, a new topology. The disparities, Richards reminded his reader, are just as powerful as the similarities, and sometimes "the peculiar modification of the tenor which the vehicle brings about is even more the work of their unlikenesses than of their likenesses" (ibid., 127). From a topological viewpoint we observe that quantity and quality are not clearly close; numbers, at least the integers, are not "like" or "unlike" each other, except as numerals, whereas shapes may or may not be similitudes, and this distinction between such divergent understandings of how to understand shape is, for me, basic.

With Renaissance science, but in the terms of ordinary language, unlikeness and disparity are a stimulus to surprising imaginative forms, and this antithetical countercurrent is a key demand that both science and art must use metaphor. Partial likeness appears everywhere when we encounter analogies and associations, but the point is to isolate the forms and shapes of difference hidden within the linkup of like and unlike. Metaphor aspires to question all semblances of iterative conformity. Thus the careworn Prince of Denmark is ironic in asking how his noble friend can be "crawling" about, like a worm. When Cleopatra is dying, the poet ironically speaks of her "strong toil of grace," which packs together many virtually contradictory meanings. Without such central antinomian figuration given by metaphor, the higher level of Coleridgean imagination will not exist, because there would be no acceptance of dis-parity, and hence no tension in the vision as a whole. When metaphoric disparity cuts into the standard dictionary meanings of words, it makes an incision into the smooth surface, an idea central to the work of the distinguished, now sadly deceased French philosopher, Gilles Deleuze, a semiotic space left open for new meanings to be folded into new and unfamiliar figures of thought. It is as if a metaphor could fold a piece of paper, drop ink into the fold, squeeze the two insides together and thence produce a New World—the making of two island surfaces to left and right. The edge is always a conceptual shoreline between two manifolds.

The Fearful Sphere

The sphere and the fold are interconnected, and somehow the sphere is the ideal shape for what is changing in historical perspective. Early Western cosmology had, of course, anticipated Sir Thomas Browne's seventeenth-century scene of "divided and distinguished worlds," a paradox caught perfectly by the much later essay of Borges, "The Fearful Sphere of Pascal," a mythic topological sketch in which the Argentinian author recounts various anamorphic reshapings of the universe as a perfect, yet also imperfect sphere. The fearful sphere has been stated in a variety of styles, of which one will serve here: Giordano Bruno's 1584 treatise, *Della causa, principio ed uno,* says that the fearful sphere is no longer a simple unitary idea; Bruno puts it this way: "we can assert with certitude that the universe is all center, or that the center of the universe is everywhere and the circumference nowhere," a conception closely following a prior medieval homiletic source, Alanus de Insulis, again quoted by Borges: "God is an intelligible sphere, whose center is everywhere and whose circumference is nowhere." Such Gnostic readings of universe as floating island attempt to express the mystery of the divine being, around whose omnipresence there seemed little room for thinking mere sizes and measurements. Instead, the divinely ordered universe is shaped to contain its own powers of containing itself, so long as it remains free to metamorphose, a transformational shaping inspired by a sacred tautology.

Ernst Cassirer notes in his classic monograph, *The Individual and the Cosmos in Renaissance Philosophy,* that Nicholas of Cusa (who also mentioned the fearful sphere) foresaw the primary attitude of the New Science, whereby men imagined themselves both in their world and simultaneously outside it. Only with such a double consciousness could Galileo have intuited that the physics of Earth would have to be of the same nature as the physics of the universe lying beyond our terrestrial biosphere. This Renaissance preference for the coincidence of opposites operates through an extreme metaphoric intensity, and it grounds not only the great art of the period, but no less the beginnings of modern science of motion in all its aspects, including Galilean relativity and the Newtonian laws of motion and gravity.

There is a metaphoric play of language (in whatever kind, be it verbal, mathematical, gestural or of any other sort) where the speaker twists the standard literal sense of the signs or the form of equations, allowing the medium of thought to guide the mind into uncharted territory. The play of language then permits strange spherical combinations of sense to guide the mind into new maps or framings of the question at issue, as if a new algebra by itself discovered an unknown set of relations between unknowns. When this happens, we may say that metaphor has animated a higher creative imagination, but we need to look briefly at the central process by which the breakout of thought occurs through the agency of language itself. This liberation from the familiar literal sense—the fundamentalist sense, we might call it—gives a formative role to metaphor and its disparities, which long ago Gustaf Stern, the Swedish linguist, discerned in a treatise he entitled *Change of Meaning.* Language is potentially a fluid medium, whose currents interfere with each other, usually under the surface of speech. Sentences end in eddies. Freely we may say that metaphor, when fully activated, can change the meanings of stereotyped definitions and expressions. That, for example, is what happens when time becomes, as Hermann Bondi said, the *mingling* fourth dimension required by the Special Theory of Relativity. As variables the four dimensions can only be mingled, which mathematics seeks to describe even when the fourth dimension, as a possible experience, is so utterly unlike and completely disparate from the three variable orientations that yield our idea of space. Gerald Holton has researched and written extensively about the mental powers involved in such counterintuitive theory, which uses ideas ultimately topological in purpose. In a collection of essays, *Einstein, History, and Other Passions,* Holton writes that the frequent resistance to science at the end of the twentieth century implies a resistance to imaginative thinking, in both the sciences and the arts. One can hardly complain about such a view, for so many scientists have told "stories" about their sudden insights. Holton has always insisted on the relevance of an almost musical inspiration for many scientists. Nevertheless, in one particular chapter, "What, Precisely, is Thinking . . . Einstein's Answer," based on research in the Einstein papers, including the private letters, there is a seemingly unimpeachable Einsteinian statement that needs to be reformulated, or rather reinterpreted.

Like all genuinely creative powers, Einstein's understanding of his own thought is ultimately too subtle to explain, and yet it clearly depended on a special sort of physical mediation, a rare embodying mode of thinking, much involved with "picturing" events. Somehow he moved from a persistently recurring image to "an instrument, a concept," but he was not happy with calling this a method of thought. The visual picturing undergoes a gradual (occasionally perhaps also a sudden) metamorphosis of its role in the larger argument. "All our thinking is of the nature of a free play with concepts," Einstein typically remarked, and he was famous for his almost musical instrumentation of thought. This is not unlike Friedrich Schiller's claims for an imaginative free play of poetic combinations, which the great Romantic considered an archetype for freedom itself, as stated in his *Letters on the Aesthetic Education of Man.* Our actual disparities change in the real world perpetually, even though the image, the icon, the picture-logo, and similar visual devices have all become ossified, even as they narrate in comics and video games, but no less in advertising generally. A ritual "ramping up of visual information" in our time has been deliberately designed to fixate the desires of any species, including humans, attuned through evolutionary development to pick up visual stimuli as shaping our sensory environment, and in this loop of responses we humans pick up visual cues for surrounding danger or promise. When repeated visual stimuli are presented to us, as in advertising, there is a danger we will stop searching our environment in any serious sense. We stop searching, and we buy the proffered commodity. *Fixated* is the psychological term for a concomitant loss of free perceptual play, the loss of what, as Edgar Wind showed in his *Pagan Mysteries in the Renaissance,* the early modern thinkers and artists lucidly named *serio ludere,* serious play, a notion more widely demonstrated for all civilized culture by the Dutch historian Johan Huizinga in *Homo Ludens.*

The ultimate model for such play may be found in poems like Ovid's *Metamorphoses,* as I have already suggested, or indeed throughout Shakespeare, but for my purpose here the model can best be linked to topological transformations of shapes. When a poet or a thinker tropes an image, that change of meaning is a bending of the original, not a destruction or even a fragmentation of it, and great flexibility of mind is required if the trope is to succeed. Strictly, for linguistic purposes, it may be that Professor Bassler is right when he suggests to me that the bending of a topological shape is not exactly the same as the crosscutting of metaphorical disparity, since the bending of a donut into a coffee cup leaves us with an essentially equivalent shape. Yet metaphors can be reversed in meaning, and my continuing concern, as will be apparent, is going to involve a special property of bending shapes to human purposes, namely the use and sense of edges, which is what metaphors enforce when they allow us to cross from one interpretation to another.

Radically we need a disparity action, as Richards claimed, for only by seeing what our vision does NOT include or convey or is even contradicted by can we more intensely perceive what it does communicate. The *discordia concors* is the basis of the final accord, and every analogy includes, if covertly, its quantum of disanalogy. Thus in Science we may need a fluid or fluent, even turbulent, notion of picturing our planetary wandering, while similarly the Arts keep on discovering their negative numbers. In a remarkable way mathematics makes discordant order possible, and the statement that time is a fourth dimension may be regarded as a mathematically inflected metaphor, an expression quite different or disparate from the literal experiential grasp of time passing such as we all have in our minds. Our sense of knowing time depends entirely on our living in a macroscopic world, paced by its "normal" sizes, and finally our perception needs to be that time both is and is not the fourth dimension, Hamlet both is and is not "crawling" on the earth—a paradox or wrenching contradiction discovered by all who invent original conceptions and ways of life. Consciousness is as much the world imagined, as it is the world perceived or analyzed or explained, and unavoidably our minds even on dull days and difficult nights are always "too much with us," as the poet said, for we are too much the children of this world, and too little of its possible illuminations.

Metaphor undermines literal sense in all sorts of ways, and the consequences of this internal reversal of sense, operating virtually under the surface, are powerful in all the arts, but they appear in all great scientific discovery as well. Here we can only suggest this expressive wealth. To repeat: Because disparity is essential for the building of imaginative forms, this antithetical countercurrent is the key feature of all so-called creative imagination in both science and arts. Likeness is everywhere when we encounter analogies and associations, but the point is to isolate the forms and shapes of difference hidden within or below the linkages of like and unlike. When Shakespeare's Cleopatra is dying, the poet's vision packs together all sorts of virtually contradictory meanings in her death, an example that may stand for the metaphorical process in general. She is at peace, and yet even that is a toiling struggle, which holds a deeper meaning of grace.

One cannot but be struck by the coincidence that Galileo and Shakespeare were born in the same year, 1564, and they both possessed the gifts of music and metaphor to the highest degree. One can only suppose that such gifts were connected with their superior powers of thinking in general, especially their independence from the run of the mill conformist ideas of their time—we are familiar with the troubles Galileo experienced at the hands of the Holy Office. Freedom of thinking in radically new ways is rarely without dangerous consequences to the thinker; one might even say that metaphor is a dangerous game—as the ancient rhetorician Demetrius once suggested—much more so than the best-selling mechanical combination of "fixities and definites" that propels any merely fanciful sales pitch. When strongly enforced ideologies prevail, the fixed fetishes required by the propagation of a faith are enlisted to gut the rich body of ordinary language and symbolism, leaving only a blank wall against exploring the actual mysteries of our unaccountably odd human lives and aspirations. We stare more and more, as we see less and less, until we become slaves of our own intolerance of ambiguity. The ideology of fact itself may promise techniques that undermine themselves, undermine their own use, their efficiency, when they fail to grasp the ambiguity of truth.

Metaphoric freedom pervades the greatest scientific advances, as with Shakespeare's dramatic inventions. Nicholas of Cusa foresaw one primary attitude of Galileo, whereby the laws of the universe might be replicated paradoxically on terra firma, our physics applying as well to the Moon or Jupiter as to the Earth. A revolution of thought occurs when thinkers in any field imagine themselves both in their directly experienced world and simultaneously outside that world. With early modern science, observation and experiment begin to acquire a studied exteriority, matching anew the views of ancient astronomers in Egypt and elsewhere. It is said that Galileo was the first person to have turned a telescope upwards to view the heavens, instead of across either land or sea. Certainly Galileo argued a different view of nature, whereby the physics of Earth would have to obey the same laws of motion as those demanded by the physics of the celestial "outer" universe lying beyond our terrestrial sphere. This Renaissance preference for the coincidence of opposites operates through an extreme metaphoric intensity, and it grounds not only the great art of the period, but no less the beginnings of modern science of motion in all its aspects, including Galilean relativity and the Newtonian laws of motion and gravity.

In his book, *Conversations with the Sphinx,* Etienne Klein analyzes several central discoveries of twentieth-century physics, such as the Uncertainty Principle or the Wave-Particle Duality, as paradoxes of both scientific logic and scientific observation. These paradoxes are cases of metaphoric disparity in Richards's sense, or mingling in Bondi's sense. To this day scientists are still searching for a good way to reconcile quantum randomness, workable only when treated with statistics, when chance collides with classical Einsteinian relativity theory, if not by stubbornly dissolving the paradoxes or simply disregarding them by blindly doing the equations. We need to think and live beyond our calculations, surely, as many discovered in the days of early modern science. On the broader macroscopic plane of the new manifold we are then forced to live with disparity, as Sir Thomas Browne perceived and Rosalie Colie wrote in the history of seventeenth-century disparities, *Paradoxia Epidemica.*

The history of paradox in logic is undeniably ancient, as will be clear from Roy Sorensen on paradox in *Philosophy and Labyrinths in the Mind,* where we are reminded that for thousands of years logicians have been slowly resolving such riddles as The Liar or Zeno's paradoxes of motion or Russell's reflexivity paradox of the set of all sets that are not members of themselves, riddles which can take many forms, including quixotic stories about village barbers that must remind us of *Don Quixote,* especially its subtle and surprisingly modern mirror stage, its Second Part. My impression is that in every paradox, and not just in the self-referential cases, there lurks a mirror image presented in one guise or another, as with The Liar, where Epimenides the Cretan announces, "All Cretans are liars," since if he lies he is telling the truth, and if he is telling the truth he lies. We cannot grasp the edge in the argument, or where the disparity is to be located; we hover between two necessary but undecidable landing fields. As Bertrand Russell discovered about self-reference, the logic is almost impossible to unravel from its skein of irresistibly asserting something about our powers of saying what we say, knowing what we know, without being stopped by a disorienting temporal sense that we are only quoting ourselves. It is as if we were condemned always to saying, "I see myself in the mirror, but if so, who am I?"—a troubled personal experience which Sigmund Freud tells us formed one starting point of his essay *On the Uncanny.*

Psychoanalysis shows that we humans are always stuck somewhere just before there was any time, at the beginning. This labyrinthine paradox appears to be the most radical kind of disparity action at work in any metaphor, as if, in order to claim a likeness or analogy between two states of things, one had to deny the claims of their similarity, by insisting on their prior unlikeness or disanalogy. One begins to wonder whether in nature the most rigorous arguments cannot help but undo their own logical entailments, whether nature does not undo logic. Logicians have worked hard to disentangle this maze, and scientists of our time continue to meet those very paradoxes that Klein has argued are perhaps the main inspiration of the most original science. Certainly paradox dominates attitudes in the arts in many ways, in modern art sometimes bringing creative imagination almost to a standstill, just the way rational cynicism readily infects our dreams of balanced skepticism.

Theory requires that we be clear about the status of the *esemplastic* (in-molding) image, to use Coleridge's term. If measuring, point by point, we reproduce the contours of any object, we have exactly graphed its image, as in a topological "graph" or in photography. Tracing the object, we get a literal imagistic double of its original body, or perhaps we should say we have used the idea of classical Newtonian mechanics to assess our imagery for its accuracy. The mechanism of this procedure belongs with the lower level Coleridge called Fancy, which depended, he argued, on the powers of memory and could only be mechanical (as distinct from "organic"). Over such a view there must be considerable debate; for example, the great critic Harold Bloom considers Fancy to be yet another mode of true Imagination, though I would reluctantly disagree. One can ask what is mechanical about the calculating "memory" of a computer-like function of an organ in the human body, with all its feedback loops. Following the early lead of Descartes (man is an automaton) and the more subtle discussion by Leibniz, recent neurological thinking has greatly developed the notion of Julien de la Mettrie, that man is a machine. Perhaps here the mechanistic and the organic meet, that is, in consciousness, which is their meeting ground. Imagination tests our assumptions about mental mechanism, however. In *Objective Knowledge* and elsewhere Karl Popper suggested that after the middle of the eighteenth century it became increasingly difficult to distinguish between living and dead matter, or between the machine and the computer and the computing human being. Mathematically skilled arithmeticians were customarily called "computers." Human computation cannot fail, however, to make errors and approximations. "Artificial intelligence" is remains a not entirely primitive figure of speech, but one asks, is this really thinking, or only recalling and following complex instructions? A sensitive use of metaphoric disparities will immediately reveal that as soon as human *consciousness* assumes the task of interpreting nature and its wiles, there is a need to accept an ambiguous view: "X is this, but not quite, or only this."

One can only ask, then, how such a machine of Coleridgean Fancy could ever envisage anything radically disparate, however miscellaneous in appearance the parts may appear. A machine can be framed to predict outcomes on the material plane; computers have what are called memories, by analogy. So far, however, without further development of quantum computing, science cannot do much more than play at the role of imagining. Our minds do that. The proposal that thinking can be done by a machine is thus a tricky one, and in our time the proposal has beguiled many who believe that thinking is a sort of complex counting process, possible now through our extraordinary computers. This view of the field of calculation was perhaps already seen by Coleridge, when he suggested in the *Biographia Literaria* that Fancy (an aspect of imaginative powers) was a "mechanical" process, a mode of computational control, a system with an Achilles Heel in the memory-bank, such as Stanley Kubrick's *Hal* displayed in the film *2001,* following Arthur C. Clarke's mysterious film script.

Yet assuming that imagination of Earth or of any other massively complex object must consciously extend its estimate of the bounds of an object under review, then the imaginative task can be said to redefine the traditional authority of memory. Imagination does perhaps lead inevitably to the Longinean sublime, but its overextension is precisely what requires either of two things: (a) an infinite amount of computer memory, as in Alan Turing's original Turing Machine, or (b) the disparity action of strong metaphor, as if disparity could achieve cybernetic control over chaos, as it were with the poet's eye "in a fine frenzy rolling." Metaphor might then animate the search for a finer sense of the rondure of our planet, which may seem to us so flat, yet is round as the human eye, even when we only see the local flatness of land and sea.

To a great extent we are what we see and especially where we see, yet our planet and its shape—that is, what we see of that shape—by virtue of enclosing us, comprehensively conditions all the powers of thought and expression—in arts, sciences, and elsewhere. In this aspect we are *where* we are, so that, parenthetically, if as the ambiguous Gertrude Stein famously once wrote, "There is no *there* there," the inhabitants of that nowhere place have no existence that we can value or they can perceive. Let's just admit that Stein was savagely spoiled, and Oakland was just too much for her sensibilities to handle. (Alison Lurie much later set a smoggy satirical novel, *The Nowhere City,* in Los Angeles.) Our human fate seems to be that there is only the shakiest equivalence between the earthly sphere we inhabit and the kind of imaginative thinking its local or global comprehension requires. The task is hard, to perceive or conceive the confusing topology of our existence, or as religions wish, to accord to intelligence a transcendent basis, as the mysterious condition of our being. Literature and fiction permit a virtually mind-altered sense of things, and in that fashion, as we find for instance in *Gravity's Rainbow,* the ineffable becomes speakable in ordinary language, but only in that mode of artistic and poetic imagination.

Perhaps it is never finally possible to evade assertions of nothingness, as if they could be said to make sense. "Darkness visible" (*Paradise Lost,* Book I, line 63) was once a Miltonic figure of thought used by the poet to describe the contradictory Stygian flames illuminating Hell; but now the cosmologist's "dark matter" makes it almost a familiar face of what is truly beyond us, an ultimate physical equivalent of the condition Milton had foreseen. Without any philosophic pretension, one could say that here metaphoric disparity has collapsed into literal statement, or perhaps the opposite is the case. We have entered the world of de Broglie, Dirac, Pauli, and quantum paradoxes.

Even more to the point, for the topological discussion, is the analogy between these disparities (which so often recall the Fall of Parity in physics, dating from the late 1950s) and the notion of a cut or an edge, incised in space or in the fabric of thought, as an act of discriminating between one perception and another close or similar to it. The critical examination of shape will counter any tendency to think we understand things if only we can measure them, for such counting opposes a deeper intuition of disparity. Metonymies, which in literature oddly rename things, abound in whatever we estimate "realistically," beginning with the nineteenth-century Naturalistic novel on which Roman Jakobson shed clear linguistic light. Such novels are always counting things, like Jane Austen's narrator referring to the amount of a gentleman's income. We need then to ask what exactly we are measuring, when we return to the original sense of "critic" and crisis, from the Greek *krinein,* meaning to judge a different state of things, or simply to diagnose a difference, a particular edge. If metaphoric disparity conveys sameness but also sharp differences in kind or quality, this will affect how we judge any imagined shape, nor can civilized people avoid the need to discriminate between differences—always without prejudice, and metaphor the mental figure of thought that most powerfully keeps us thinking about sameness and difference, about the scientist's symmetry and dissimilarity. Without metaphors working in our thoughts, it is unlikely that humans would understand what is incongruous, for as Hermann Weyl wrote in his book, *Symmetry,* significant shape is given by *automorphisms,* "defining them with Leibniz as those transformations which leave the structure of space unchanged" (1952, p. 42). In this sense disparity action is the key to topological invariance, as it was in Ovid's *Metamorphoses.* It is as if a hidden continuity comes from reshaped opposites, flows from dislodged, relocated fragments.

There is an irony in the fact that disparity can produce a larger "implicate order," to use David Bohm's thought in *Wholeness and the Implicate Order* and in essays on creativity. Bohm shared the currently infrequent view that art and science need to cooperate, *as modes of thinking.* They should not be systematically cut adrift from each other, which commerce always seeks to do, by turning science into technology and the arts into "mere decoration." A peculiar isolation of thought and feeling may too easily arise, denying our sphere by rejecting its potential for isolation. As we proceed, our task is to deepen our sense of the whole Earth as in fact a cosmic island, by asserting continuity and tradition and change, all together and all at once.

IV

Euler Discovers the First Edge

Let us ask more questions about the way topology intersects with our lives, minds, and especially language, if language is understood in a Galilean fashion, reaching beyond ordinary speech to include a mathematical symbolism, but still connected to written treatises such as Galileo's *Celestial Messenger* or his later *Dialogue on the Two World Systems.* With the Königsberg Bridges we have seen that vital changes of critical positioning—the sites where we locate points of arrival and departure, and the sequences of stopovers (some more efficacious than others) on any journey—were the subject of the original Leibnizian dream, a mathematics of position. On that view the structure Hermann Weyl mentioned in his book *Symmetry* is not that of abstracted symmetrical order alone; crossing a bridge is also a "real" motion that defines constraints upon any next crossing and in that way determines the graph—the line drawing, in effect—of a larger assemblage of crossings, let us say their combination into a single Eulerian Path. It is as if Euler asked himself what principle is involved when we move from one point in space to another, or as we say, we try to find our way. This is a matter of thinking more exactly about pathways, about what it means for a bridge to cross a river at an angle, reaching the other side, touching the surface of dry land beyond its ends. We speak of the edge of the river, but only before we attempt to cross it. Somehow every bridge is also an edge, but is an edge that starts from another edge, the bank of our theoretical river. So we must understand edges.

Amazingly, no mathematician before Euler had ever defined the shape-making role of the cutting edges, which are basic to the Königsberg pathways. Before 1750, when Euler published his second great topological paper, no mathematician had even *named* geometric edges, let alone defined them, almost as if their form and function were too obvious even to merit distinct geometric denomination. David Richeson's recent study, *Euler's Gem: The Polyhedron Formula and the Birth of Topology* notes the anomaly, when Richeson says that on November 14, 1750, the newspaper should have used a banner headline announcing that a famous mathematician had discovered the edge of a polyhedron. Euler had recently written to his friend Christian Goldbach (famous for his Conjecture) that he was thinking about "the junctures where two faces come together along their sides, which, for lack of a better term, I call 'edges'" (2008, p. 63). This lacuna suggests that because topology did not yet exist in Antiquity, as the science of shapes and their continuous deforming, it was fundamentally not concerned with measuring sizes and distances, unlike geometry. Geometry assumes a precise definition of the form of such a line, which it could treat in axiomatic fashion, as the shortest distance between two points, that is, a measurable distance. In Euclid the line connecting any two points was a traced or drawn line, which in his Greek was called *gramme,* from which we get our words "graph" and "graphic." A straight line in Euclid is then simply a well-drawn or well-made line *(eutheia gramme),* but Euclid was always thinking abstractly in ideal geometric terms, with a very different sense of what it means to walk over a bridge, when we assume a vision of physical, spatial quality rather than angular quantity.

Our topological account now calls for its second act. Having with the Bridges of Königsberg already invented topology as a new kind of networking combinatorial mathematics, Leonhard Euler had turned to other problems—his complete works are estimated to fill over seventy volumes.

At some point in time he switched to analyzing the Platonic geometric solids, work from which he achieved another conceptual triumph, his next fundamental discovery, which is called *Euler's Polyhedron Theorem.* Of those rare Platonic solids there exist only five distinct types: four-sided tetrahedron, six-sided cube, eight-sided octahedron, twelve-sided dodecahedron, and twenty-sided icosahedron. These are unique shapes. Having different numbers of divided surface areas, they are called polyhedral. The sides are all plane surfaces, as for example an actual pyramid, which can be reduced geometrically to the first Platonic solid, the tetrahedron. Partly because Plato had discussed this group of shapes in his *Timaeus,* which asks about the shape of the universe, they fascinated Johannes Kepler, who thought they contained the structural principles of cosmic order. The component shapes of their plane faces are regular, for example triangles, hexagons, or pentagons. The terms *face, surface,* and *side* all here mean the same thing, applying similarly to the rectangular sides of a cube or the triangular sides of a tetrahedron.

In Euler's analysis the startling novelty of his earlier 1736 topological bridging discovery was surely not lost on him. He had seen that an ideal emptiness, an abstract network of space, could be shaped as graph of place, an abstract line drawing of the component positions of sequential movements. The single position of a bridge-end could be assembled into a composite system of bridge-ends in the plural. To our knowledge he did not actually draw a graphic schema of the Bridges—that method came much later—but clearly he pictured such an outline of continuous tracing, and in due course other mathematicians drew reductive graphs, which we call networks, composed of lines and nodal points—something like a map of air-traffic paths between stopovers and destinations, showing the implications of the sequencing whose controls Euler had simply imagined.

In early 1750 he was pursuing his topological approach, as before committed to avoid or evade measurements of size. Contemplating the Platonic solids, he saw that these rare mathematical objects all shared a defining set of interconnected features: namely, they all have vertices (as a pyramid has one topmost point, a cube eight points or corners, a sphere has no such points at all), they all have edges connecting the separate vertices, and they all thereby have sides, or faces. Every edge separates each face or side, for example, a field extending away from a stone dyke or wooden fence, perhaps running from a second boundary which marks the edge of yet another field. This is the image Robert Frost developed in his famous poem, "Mending Wall," and indeed imageries of edge and position abound in that poet's work. Old New England common sense might find the fact of rock wall edges unremarkable, but theory could earlier ask more abstract questions.

If in Antiquity no mathematician had focused deliberately on the shape-making role of edges, it is as if geometers understood edges, but were too familiar with them to name them as such. Even Descartes, who had anticipated Euler's work regarding essential polyhedral properties, appears not to have noticed that edges were special, that they implied a fundamental statement of topological invariance. In *Euler's Gem* David Richeson explores a wealth of consequences as the original mathematics began to develop, while the topology centers on its originating polyhedron theorem. At issue, as with topology in general, is the search for an invariant in regard to shape, and we are not disappointed. The principle sounds transparent enough, but it helps to see how it describes a material object like a pyramid, whose essential shaping properties (V, E, F) are listed and interrelated according to the *Polyhedron Theorem:*

> A polyhedron with so many (V) vertices, so many (E) edges, and so many (F) faces satisfies the equation $V - E + F = 2$, and this sum is always the same, a constant, no matter how many sides the shape possesses. $V - E + F$ will always yield the number 2.

Strikingly, the vertices or points ending or starting any angled side are zero-dimensional points, the edges are one-dimensional lines, and the sides are plane two-dimensional polygons. What is so striking about Euler's polyhedral theorem is its invariant result, the number 2, since a scarcely imaginable number and variety of different objects, including the sphere, end up with this same number 2, no matter how large or small or variously shaped these solid objects may be. There is a remarkable consequence of the theorem: whenever solids are altered in shape, an invariant central equation comes into play, while a similar but more complex numerical invariant result also appears when objects are pierced with holes.

Boring a hole right through a complete shape such as a spherical ball stops that shape from being simply connected, a basic property which it possesses, so long as it is not cut in two or is pierced by a hole. These latter intrusions cancel the integrity of the original shape. Thus it matters a lot whether the connexity is preserved on the surface of the fundamental shape. With a donut-shaped torus the hole in the center prevents any simple connectedness, because any loop drawn on the surface of the torus around its twin curvatures fails the test: such a loop cannot be shrunk to a single point without leaving the surface of the shape.

This difficult picture of an index to completeness is less important for my purposes than noticing that if a shape is to preserve a primal integrity, its "soul," as topologists sometimes say, it must be simply connected so as to produce continuous flow from one point to another to another. A space (like a material object) is connected, then, if it is all of a piece, and we have already seen that with the Bridges of Königsberg, the whole site was treated as one potentially connected space whose internal edges (the bridges) required analysis. What then, we ask, constitutes the whole of anything, or what would wreck that wholeness? Variety itself is not a home wrecker, we find, as certain cultures also find with human marriages! It is as if the simplicity of the connection makes possible a continuous equivalence or the nearby *homeomorphism* of points shared by different-looking shapes, such as potatoes and pyramids. (The animated motions of a coffee cup turning into a donut, in the *Wikipedia* article on homeomorphism, strikingly illustrate how space itself may without fundamental change be molded and remolded.) The invariance at the heart of the change is clearly what matters most for this homeomorphism, and in art we often notice this effect, for example in the use of invariant human-body or acoustic-guitar shapes to anchor the Analytic Cubist paintings of Braque and Picasso.

Such paintings indicate what is going on here: *In Pursuit of the Unknown* by Professor Stewart shows how any particular solid can be reduced in complexity, by cutting away faces, edges, or vertices, and "these changes will not alter the number F - E +V [his favored order for V - E + F] provided that every time you get rid of a face you also remove an edge, or every time you get rid of a vertex you also remove an edge" (2013, p. 92). The fundamental shape remaining after a series of such self-canceling surgeries is finished is a single vertex on top of "an otherwise featureless sphere." A complicated solid has become a sphere. But as Professor Stewart goes on to show, the donut-shaped *torus,* with a hole in the middle, will not yield to this transformation; it cannot be shaved into a sphere, for the hole prevents that operation. The attempt will not produce the magic number 2, a failure proving that the sphere and *torus* are topologically different shapes. The vertices, edges, and sides combine as an invariant group or association, and this invariance has meaning: it implies that with solid bodies, such as the earth, the fluctuations of shape do not alter the essential form. From our present point of view this is a good distinction, since I want to stress the importance of the two-dimensional place where humans live, which is not the same thing as the three-dimensional space *in which* our terrestrial place is located. To construct our sphere ideally requires only two dimensions, but to convey how that body is variegated with ups and downs, it is much easier to visualize its three-dimensional situation in space. Once again we are reminded that our planet is critically endowed with a certain shape, which involves our entire mode of life, that is, our being alive on its essentially two-dimensional surface. This dynamic situation of our being alive, along with myriad other living creatures, larger and smaller than us, demands the fairly exact planetary form that Earth possesses.

The Sphere and the Edge

While our planet Earth is shaped like a very rough version of the ideal sphere, in actual fact it is an oblate spheroid, squashed at top and bottom, putting on weight in the equatorial region, and full of surface bumps and declivities. Despite these imperfections of form, the sphere has two faces (inside and outside), these faces have no boundary, and *ideally* our Earth has no edges.

It is this second property that is most curious, when thinking about the massive manifold of the earth's surface, inhabited by living creatures at all macro- and even microscopic levels. In ideal terms our planet fits the Polyhedron Theorem, which we recall runs as follows: Vertices = zero; Edges = zero; Faces (sides) = 2. The Earth fits the Theorem, $V - E + F = 2$, and at a high level of abstraction this invariance is as true of our sphere as for the cube or the pyramid, or any others among an infinite number of polyhedral forms, *all of which can reduce to the fundamental form, the sphere.* This holds true so long as we do not cut the shape in two and then try to sew its halves together again. On a smaller scale, objects with large numbers of faces obey the same law, and if Earth were shaped like a gigantic soccer ball, the result would be the same. The quotient on the right-hand side of the equal sign is always 2, no matter how many faces the solid convex object possesses. This might seem impossible, yet it is the case, for all such examples are reducible to a sphere.

Strangely, the Theorem tells us that, despite first assumptions, the spherical shape of Earth requires that it have no edges. The phenomenology of Merleau-Ponty finds such things unsurprising, for in his view, as perceptions, round objects like Earth have a quality of fullness, like the pleroma of Being itself. Gaston Bachelard observed the same thing in the penultimate chapter, "Intimate Immensity," of his 1958 book, *The Poetics of Space.* For Bachelard, inside and outside are phenomenological as well as geometrical imaginings, showing the roundness of the sphere as if it were an approach to unified order. Cousin to the sphere, the torus, shaped by the hole in the donut, also has no edges and when closely studied shares this property of phenomenological fullness, a roundness which, as one poet wrote, invites a caress.

All other common polyhedra such as pyramids do have edges, making the sphere and torus (along with pretzels) seem rare shapes. While we do not actually live our geometry, using it only for measuring purposes, in our lived phenomenological and artistic context the edge itself, in fact and idea, makes possible the practice of drawing distinctions that go beyond simply measuring things. The terrestrial sphere (Shakespeare, punning, called it "the great Globe itself") with its two sides and no edges project a cosmic *discordia concors,* a kind of final disparity action in all metaphor, certainly a governing material source of paradox. The edge then demarcates the place where distinctions need to be carefully drawn, which the ideal sphere cannot permit, because in principle it has no edges, but which real-world life and observation demands, even so. In that sense the sphere in itself demands practical mysteries. Yet our thoughts will escape this darkness, by using metaphorical disparities as the archetype of the discriminating edge.

Having no edges, the sphere defining the shape of Earth formally demands an extreme disparity action, because we must correct primitive common sense, which seems *wrongly* to demand divisions of the planet into clear-cut flat pieces with reachable corners. On the contrary our explorations can work well only because—for we humans are physically not FlatEarthers—we can traverse the surface of earth, by land or sea, without the ungrounded fear of falling off. If things go wrong, as some FlatEarthers have believed they might, one could fall "over the edge." In various ways Jeffrey Weeks's textbook, *The Shape of Space,* when it retells the story of Edwin Abbott's classic fantasy, *Flatland,* shows how difficult it would be for a flatlander to avoid such a catastrophe, as René Thom might call it in a technical sense. The fear would be rational, given the topology of a perfectly flat planet, but in fact most humans live in the perspective of nearby flat areas, so gradually does the earth's curvature show itself. Many barriers also present themselves to humans, as animals also know their territories. Tribal peoples living on land have always known and used the material edges branching everywhere, a crazed riot of boundaries, if only natural barriers like rivers and escarpments.

Obeying the tendency to stay home, the problem of edges may easily disappear, as well it may in urban layouts. Thinking about tribal societies, on the other hand, we are not surprised that Abbott's classic parable uses the geometry of the first four dimensions to analyze the social separations familiar in the Victorian British class structure, a virtually private tribal structure, whose boundaries established archetypal social edges to maintain acceptable boundaries of knowledge and social mobility. For Britons the actual edgeless sphere was located *out there,* somewhere beyond; the British Empire was so large and so easy to spot all over the globe, that one could all too easily believe the Empire itself has no edges. That Empire, surely the last of its kind, was its own Globe Theater.

It often appears that theory precedes knowledge of fact, and in ancient times it was surely hard for humans to see that, in theory at least, spheres have no edges—this being a theoretical, ideal object—even though any child would discern the difficulty, while searching for the edge of a round ball or wondering about a soap bubble before it bursts. Dramatically enough the child would already have to know what an edge really is. On the other hand the modern ecologist discovers natural edges everywhere on the surface of our planet, which is just as imperfect topologically and geometrically as was the Moon when Galileo's telescope first allowed him to sketch its rough-hewn mountains and declivities, much to the distress of many Catholic astronomers who had been taught that the Moon should display a translucent crystalline perfection pleasing to God. The mountainous surface of the actual Moon was a threat to orthodox beliefs concerning religious boundaries and natural limits.

Modern science neutralized this doctrinal ambivalence, but something like it persists to this day. In all such cases a psychological conflict arises between an ideal abstraction of boundaries and the opposed materiality of the rough edges of real life, so that our metaphoric and imaginative disparities must be sought in the diversity of life and in its miscellaneous irregularities of actual setting, all of which involve actual edges, and then we have to use disparities in a reasonable way, as if they were borders between countries. Too often we humans are locked in a conflict between the actual and the ideal, no matter how hard we try to observe our conscious vagaries of insight, or on the contrary try to avoid imagining what lies beyond the immediate horizon.

As happens so often we discover that humans are troubled by imperfections, as soon as they begin to think. Evaluating the legacy of Riemannian curved space, Professor O'Shea remarks on our chief terrestrial situation that "the surface of our world is topologically (that is, can be put in continuous one-to-one correspondence with) a sphere, but it is not a constant curvature surface. It is not perfectly round—it is flattened at the poles, and there are bumps (that is, mountains and valleys) on it" (2007, p. 100). Our Earth is like the Moon, quite imperfect as a sphere, and this would include imperfections of edges, we should note. It is as if our most precise sciences show what art has always shown, that we are surrounded by imperfections.

Professor O'Shea quotes Riemann's translator, the brilliant mathematician William Kingdon Clifford (1845–1879), who described surfaces which are "on average flat," although there would frequently be found "little hills" dotting such a surface (ibid., pp. 100–101). Everywhere in nature indeed we find such irregularities, whose shapes are determined by edges of one sort or another, and with a lexicon of names our common languages reflect this fact. Irregularities that slightly deform larger seemingly continuous shapes surround us everywhere, which seems to provide an almost cosmic model for the true meaning of physical "shape." With great prescience Clifford wrote a theoretical memo, "On the Space Theory of Matter," (also reprinted in Newman, *The World of Mathematics*, Volume I). Which states that

> (1) . . . small portions of space are in fact of a nature analogous to little hills on a surface which is on the average flat; namely, that the ordinary laws of geometry are not valid in them.
>
> (2) That this property of being curved or distorted is continually being passed on from one portion of space to another after the manner of a wave.
>
> (3) That this variation of the curvature of space is what really happens in that phenomenon which we call the motion of matter, whether ponderable or ethereal.

These notes are virtually archetypal for our present worldview. Clifford was anticipating much later notions of quantum wave-particle shapes, energy, and matter being the results of the curvature of space, as if the solidness of matter were a macroscopic illusion produced by the compactness of the incredibly small wave motions captured in Schrödinger's quantum-mechanical equation. Given where we humans live, it seems, we have no great reason to believe in knowable perfections. Seeking for exact limits and edges, we find only approximations of their form.

That is exactly the point—that edges alert us, they do not strictly define us. The idea of anything being "hard-edged" is of course questionable on many levels of material and mental existence, and Nature obeys this law of approximation, so well described recently by the British computer scientist, Kees van Deemter, in his book *Not Exactly: In Praise of Vagueness.* Nature appears built from "vague objects," but like other terms and words involving motions and emotions, "edge" has acquired many secondary incisive connotations, as in the words "edgy," "cutting edge," Hamlet's "edge of doom," "over the edge," or other phrases implying intense desire or unbounded success. Fundamentally, of course, the term is spatial in meaning, although the emotions attached to edges are often revealing, quite beyond the field of pure dimensionality. A similar linguistic phenomenon is familiar to us from words like "set" or "drive" or "static" or "shift," where not surprisingly motion and emotion are linked by associations and semantic overtones.

For planetary ecological purposes, however, the chief interest of the idea of edge is that it illuminates our blind spots, our instinctive belief that somehow the Earth ought to have a more perfect shape. A mathematician friend writes me that "this fundamental conception of edge is ultimately as paradoxical as the disparity action of metaphor, and something only partially captured by attempts to render edges fully mathematical." The sphere is a form that seems so simple as to get nowhere, which presses us to find some ultimate mystery, calling us to chase the horizon. We seem to be pursuing curvatures in every direction, and we cannot help but feel that sphere and edge shape the very mode of our existence. "News from a foreign country came," wrote the seventeenth-century mystical poet, Thomas Traherne, and imagination always looks beyond those strange new seas of thought, as the Elizabethans and their inheritors the Romantics always insisted. Yet "beyond" almost seems the wrong word here, because it should include thoughts of depth, meaning "within"—it should include in one single process the connectedness both *between* and *within*—yet therein sits one major earthly paradox. Loving or hating our home territory is a natural human desire, but this instinct toward the familiar tends to favor an uncritical cast of mind, and we daily see how fanatically some will maintain their so-called beliefs. The edge is the beginning of the difference between here and there, a disparity without which no human could learn to discriminate either real or imagined differences; nor could we imagine Earth, which in principle has no edges, and once again the abstract and the concrete come into close proximity. This meeting eventually should, when examined in depth, along with the distinction between shape and scale, lead to ethical questions of action and decision, to which, it might be argued, the topology has only a qualitative or impressionistic linkage, unless we agree that ethics implies a continuous or customary mode of behavior. It follows that in ethical terms we return to our earlier question: what are we seeking in this fabrication of edges, with or without their topological significance? The question may be rephrased as follows: What are the persistent ecological consequences of this making and unmaking of edges? Previous pages have suggested more than one immediate use, for example, material edges following designs of bridges or wagons. My answer can only be: Edges are the great instrument of human creativity, commonly found when producing artifacts. But they come in all sizes and shapes, often in the curvatures of thought and feeling, and they always mark moments or phases of differentiation. Somehow we have to balance their divisive function, treating them as the method of neutral values and concepts. For us humans, for us animals and for the registered wild animals we seem hell-bent to destroy, edges are critical precisely because in principle we all live on a sphere, which has no edges, a surface that needs to acquire conceptual boundaries for practical reasons, which can only happen through human intervention.

Our immediate task will be to engage in what seems a contradiction, namely that our differences as perceived and understood are the best if not the only disparate method for reaching a functioning larger clarity of purpose. Holism only makes sense if it is cut up into the right pieces, so for that we must use incision. As we shall see, if this does not happen, we humans will defy all Natural Edges and will misuse all humanly devised Constructed Edges, *their erasure leading to collapse of the whole.* In practical or ethical terms we ask what link connects such material edges with the less material edges of thought and cultural difference. Ethical behavior is not a one-shot affair; it involves a way of life, for human life cannot be engineered, despite all suburban appearances.

The larger pattern, like David Bohm's *implicate order,* is what counts, to revert to the language of numbers for a moment, as the numbers meet the drama of meaning, the drama of scale. This turn or return in our thought is always going to summon ideas of the imagination, and therefore, since metaphor is at the heart of imagining things, our fundamental topological concern for edges remains central. We need what I have elsewhere called a liminal vision of passage, which in turn can only be uttered in what grammarians call the Middle Voice, which is neither active nor passive, but oscillates across a middle movement of thought. Speaking for an earlier American life with its faulted aspirations and its high hopes, Walt Whitman's *Sundown Poem,* "Crossing Brooklyn Ferry," captures the feel of this transitional worldview, where sooner or later the sun must set, but where the sun also rises. Every boundary is intended to be crossed, sooner or later, however dangerous the crossing.

It follows that if every edge implies intelligent, humane discrimination, a sense of differences and a fair judgment of their importance, it follows that action at the boundary is a special kind of awareness in the taking of an action. At the very least we may call this an awareness of what lies or may lie "on the other side." This sounds much like an intellectual demand, something a utopian philosopher or precious interpreter might vacuously desire—to voice an unkind prejudice against conjecture. But it is no less a pragmatic concern and, even on the further reaches of thought and feeling, a question about what lies beyond common experience. Such finally is the virtually Aristotelian basis of our seeking an ethic based on awareness of the middle conditions lying between extreme positions, for the sense of edge, of actively discriminating where lies the boundary and wherein lies its purpose, is exactly what is called for when we want to move into the massive area shaped by the middle ground between extremes. My view is that if we understand the topology, or perhaps only the layout of transforming places, we will come to neutralize any prejudicial idea that "discrimination" is inevitably a hostile game of self-proclaimed superiority scorning supposed inferiority. We will also defend the use of edge against those who would tribalize it, in hopes of kindness. In this way an ethics of scale might try to illuminate the journey toward a neutral scientific opposition to waging wars of revenge and resentment. The imprisoning use of boundary is not as clear as we would like, but it certainly involves making war and using a boundary in order to prepare for war behind its Siegfried or Maginot Lines and similar defenses. In today's robotic aerial warfare, of course, such Lines belong with thousands of previous casualties of progress. The gates of the city are no longer safe, neither open nor closed. With the advent of cyberattacks this is even more the case, and the definition of war itself has changed radically, since information is no longer an easy thing to keep secret. The Trojan Horse has returned. The usual dialectical principles apply and once more in human history we find that every new weapon (every new medium, Marshall McLuhan would have said) begets its counter weapon and counter medium. Nor is the isolationist ethos of tribal organization a sufficient means to make countermoves, the necessary leaps and bounds—the sphere is simply too large and round, while the sphere implies continuous rotating influences on cultural development.

V

Vico and the Cycles of Human History

The human experience of time and space engages our minds on many levels, and yet, to simplify for a moment, recall that guarding the gates of humanistic inquiry there will always be a sign that says: *Note your conception of historical change,* for therein lies the ultimate edge—the long-term limit of human experience. As we have seen, however, topology and other mathematical fields appear timeless in their application to nature and form, so that an adjustment of lens is required if we are to relate our thoughts to our *experience* of such thoughts.

My aim now is to suggest a change of paradigm, such that in the lives we actually live, with every possible variety therein, we question the blanket ideological belief that technical advances lead to an *unquestionable* progress. Unintended consequences leave twenty-first-century societies awash in doubt about such advances and their side effects, despite many legitimate claims. What we often call blind progress is just that; not always, of course, but the world population has become more attentive to risk. Talking this way to an engineer will sound hopelessly backward, but the fear has considerable merit, because we are mortal beings; we do not actually live in any progressively linear time-sequence, except in the most provocative and irrelevant perspective. Space programs aimed at finding life on Mars or determining the width of our galaxy in light-years provide no data I can relate to my next birthday, assuming there is one. The "life" and the "width" do not belong to our macroscopic scale.

Even climate change, which occurs owing to a linear progression on ecological levels and scales, can only be experienced on a nonlinear basis. Our experience of such changes comes at us in periodic packages of small-scale cyclical change, we being far less cosmic than we often like to believe. As Pascal observed, the vast spaces of the universe are mainly important to us because their magnitudes frighten us. We live according to fundamental cycles of biotic fact, and the cyclical model arises from our being neither cosmic nor at the opposite end of the scale like creatures the size of cells, but like other animals we have sizes and physical powers on large macroscopic scales. By the same token, if we descend the ladder of scale sufficiently, we begin to approach living creatures whose consciousness and language differ so radically from ours that we can hardly, without triviality, consider them to have what we call a "history." Entomologists will challenge such a statement, and so be it; they have their idea of community. On an even smaller scale so also have the scientists of cellular life-forms. Yet let us not confuse the smallest or largest communications—signals and receivers—with what we rightly call ordinary language.

The belief that signaling implies linearity needs to be questioned. If, as I intend in the present comments, one seeks a nonlinear model, much depends on what one thinks conversation amounts to, but at the very least it appears that humanly interactive history—the account and structure of events which humans believe they or others have experienced—must start from a cyclical mode of understanding, only secondarily linking that periodic mode with a large-scale linearity of form as well, a pattern usually called Progress. To grasp this combination we apply various notions of shape to the temporal frame in which we believe our thoughts have evolved.

Let us then place timeless purely mathematical topology in the lived framework of time. Despite Euclidian geometry and its heirs, despite the unchanging character of their pure form, our actual experience of events always subjects the Platonic idea to the winds and weather of historical change, or rather, it is this conflict between the ideal and actual, implying change, which inspires mankind and therefore needs temporal clarification. Beyond the plane of experience, furthermore, objective fact casts doubt on all idealized accounts of human behavior, and yet history records utterly irrational human behavior at all periods. Just think of the persistence of ritual tribal warfare, which in recent centuries shows the marks of change only in its deadly technology. Weapons sell better than any other commodity in our world. A *theorist* of history can only question the permanence of abstract linear order, and I am reminded that topology itself is not immune to this threat of misplaced concreteness. The dazzling advances of mathematics play a large role in regard to these uncertainties, and one might say, for instance, that there appeared to be almost too many great mathematicians in the eighteenth century, as Ernst Cassirer suggests in his *Philosophy of the Enlightenment.* Cassirer reminds us that Diderot's essay *On the Interpretation of Nature* (1754) held that a turning point was about to be reached, when the author hinted that "in less than a century we shall not have three great geometers left in Europe. This science will very soon come to a standstill where Bernoullis, Eulers, Maupertuis, Clairants, Fontaines, d'Alemberts, and La Granges will have left it. They will have erected the [pillars] of Hercules. We shall not go beyond that point." So much for the great encyclopedist's version of a paradigm shift!

However questionable was his prophecy, there was nonetheless a genuine sentiment behind the prediction of Diderot, who believed that among the various natural sciences mathematics would not be able to maintain sole mastery much longer. Cassirer's question had long before been anticipated by Francis Bacon, if not even earlier by William of Occam, a question Diderot enhanced by saying that "the abstract sciences have occupied the best minds for too long. Words have multiplied endlessly, but factual knowledge has lagged behind. . . . *Yet the true wealth of philosophy consists in facts, of whatever kind they may be.*" Nineteenth-century philology and materialist positivism were about to change all this. Diderot was a clarion of the oncoming fact-finding linear Progress, although he could not foresee a torrent of equally new abstract mathematical knowledge, for example with Riemann or the equations of Clerk-Maxwell. The arc of these and other intellectual rivalries, with their great developments in modern European history, amount to a world of extremely varied thought whose history determines our own, even now in the twenty-first century.

Such competition is the intellectual climate subtending Euler's invention of topology, a mathematics casting doubt on the frozen language of ideal Platonic forms. For centuries, to give but one main example, Euclid's authority was so great that the interior angles of triangles had appeared inevitably to sum to 180°. This sum appeared a Platonic truth, an eternal verity. Yet topology suggests otherwise, and the masters of curved space, Lobachevsky, Bolyai, and especially Riemann, in the nineteenth century were to discover that Euclidean convictions could be questioned, as not in accord with curved space. Theirs was deep mathematical discovery, yet I aspire to follow their example on a more common level of thought.

My current thinking is not so dissimilar, in the sense that as historians we too may have to allow radical transformations, historical change itself being understood as "curved," and we then need to consider various ways in which our notion of history is framed. If Euclid's measuring *Elements* shaped geometric theorems for millennia, we may ask, by analogy, what universal contours of the writing and understanding of history have persisted over time, to guide human inquiry into the ideas of past, present, and future events, gathering those ideas into a single model of their meaning, relevance, and dynamic. This is critical to my own account of the topology of imagination, for by shaping our historical frames of reference, we create the basis and indeed the entire possibility of asking: whose topology is it, in any case? Topology on a human scale is not the same as cosmological uses of the method, as the most casual glance into astronomy will reveal to the surfer of the Internet. It is the human scale that concerns me here, and two universal models for history at once call for brief comment, the second of which, a cyclical model, is what matters most for my topological account.

We are led to consider the shaping of physical theory as congruent with different historical models. In the West probably the most familiar theory of history—linear in shape, linked always to "the idea of progress"—builds its claims largely on scientific or technological advance, but also on improving humane values, improving in principle if not always in actual fact. Like the mythical Arrow of Time this progressive theory aims to shift its gaze always ahead, until Progress strikes an ideal target. Moderately conceived, its motto might be: *Progress Not Perfection.* Scientists constantly tell us that although a certain problem has not been solved, it soon or eventually will be.

Sharing the same concern for modest neutrality, but quite differently oriented, a very different theory of history may be reexamined for its apparently non-progressive modes of innovation and progressive advance, which occur in the context of *cyclical patterns of recurrent cultural behavior.* The issue is almost one of class warfare, because the theory of linear progress, as opposed to cyclical recurrence, intends to leave many people behind, as not intelligent or aggressive enough. On the other hand, if a critical debate opens upon questions of scale, such as a general quest for equity, my cyclical second model has its special force, for it resists unquestioning dogmas surrounding the sublime immensity and the remote grandeur of cosmological discovery. In the context of ordinary human life we scale our dreams down to the level of earthly existence, and although dreams are personal and in that sense authentic, their localization of perspective carries with it no obvious universal rule of size, small or large. My later comments on horizon as an edge specifically address this question of the unavailable, not to say ineffable, size of thought, and usually we must be content to map possible futures and probable readings of past and present: topology is important here, because it provides a mathematical approach for at least intuiting the outlines of these questions.

Topology, of course, is used equally to estimate shapes on the largest or smallest scale in physics, where instead of a mathematics of calculated magnitudes, we may choose to focus on the middle range, the macroscopic scale on which humans live. We commonly measure things and events lying within this middle scale, to be sure, but if we concentrate on the "great globe itself," where we are only players, we discover that this scene so sharply reduces the cosmos that we are obliged *to reduce its mode of history as well.* By "mode" I mean that mankind is the measure of everything, but also that *modus as measure* means a wave of recurrent repetition, where the size of an event limits the way it fits into larger historical patterns over time.

Shrinking our patterns of repetition down to size requires a cyclical theory of history, a theory most richly (if eccentrically) developed by one independent Enlightenment scholar and thinker, Giambattista Vico (1668–1744). His theory of history supports the present account, because with almost mythic force it embodies a theory of permanence in change, like topology. Vico was a latter-day pioneer of the strengths of a nonlinear mode of history, a cyclical model by and large, although he was ultimately concerned with the future destiny of the human species.

Vico's Model of Historical Change

In the year 1744 there was published in Naples the final edition of Giambattista Vico's remarkable book, *The New Science,* which argued that human culture evolves (and in that vague sense changes linearly) according to a vast sequence of recurrent cycles, obeying a model of historical change derived from a virtually mythic reading of the *Elements* of Euclid. In his book, altered in various editions, Vico sought a schematic model for the vastness and variety of historical data, and in that quest he decided that history could be seen as shaped in some fashion. A moment's glance reveals that Vico took myths and etymologies almost too seriously, but from ancient stories and poems and religious traditions, so far as one could discern them, he accepted the "forward" march of mankind as being enriched within long epochs, by a contrapuntal periodic process of reflection and cultural consolidation.

Given a broad enough chronological span of time, each age succeeding another, there was clearly a gradual unfolding of some kind—progressive perhaps, but not necessarily. Each period reflects its own sense of implicate order. Despite Francis Bacon's early seventeenth-century works on the "Advancement of Learning" (John Donne's "newe" Copernican "philosophie" being an example) and despite all refinements and reformations and later enlightenments, the turbulence of historical change remained a constant, always turning and returning in a loose system of cyclical revolution. Such patterns of change intermittently followed a larger Vichian schema of arriving at new historical moments or periods, with differing paradigms of thought depending upon something like a wheel moving forward, a slowly circling progress of social organization, as from tribal to civic life. Veering away from his earlier allegiance to Cartesian principles, which understandably were sharply antihistorical, Vico the anthropological theorist was moved to support Baconian ideas of science. In this regard, if we discount the special technical innovations of early modern science, we find that Vico's cycles are closer to Thomas Kuhn's paradigms of change as described in the *Structure of Scientific Revolutions* than we might at first suppose, subject to much the same sort of critique Kuhn himself received. Meanwhile, paradoxically, since Vico himself wanted a new science, there is at times a strong resemblance between the Vichian model of history and the archetypal and finally romantic "Myth of the Eternal Return," which appears in the oldest known religious thought and was brought back into modern philosophy by Friedrich Nietzsche. One suspects that history makes no sense unless it is shown to have certain repeated human motifs, and in all such accounts the issue is one of seeing history as possessing something like a narrative shape, which, like an old ballad, builds in periods of completeness.

Classical Roman histories and the much later scientific ideas of the late seventeenth century were, however, inevitably the chief resource for Vico, a trained humanist investigator. His career reminds us again that the Greek term *historia,* as used, let's say, by Herodotus in his account of the Persian Wars—Cicero called him the "Father of History"—simply (or not so simply) meant *inquiry.* Numerous books have been written concerning the nature of historical knowledge, though not as many as we need, concerning historical ignorance.

From Frank Manuel's 1965 lectures, *Shapes of Philosophical History,* we learn that Renaissance authors depended heavily on cyclical models descending from ancient texts like those described by Jean Seznec or Anthony Grafton, and as early modern seventeenth-century conflicts over scientific foundations continued into the Enlightenment, it became for us more important than ever that an inherited theory of recurrent historical periods, however archaic it might seem, might still appear centrally in the cyclical argument of *The New Science.* It is at least possible that even if periods as identical historical sequences cannot recur, not even roughly speaking, their spirit or ideals might well recur, approximating similar shapes of purpose. Vichian cyclicality revolves around three recurrent phases or shapes of historical change. Throughout Gentile history he finds a universal and natural succession of governmental forms issuing from a radical arithmetic: from the One, which yields monarchy; to the Many, which yields aristocratic rule; to the All, which yields democracy. Noting that Vico accepts an ancient Aristotelian wisdom as a guide, Frank Manuel further accords this wisdom a new turn, a psychological picture of the mind that leads finally to cultural history, On Manuel's view of the shapes of philosophically defined history, the three stages of cyclical change share, albeit with differing cultural styles, an evolution transcending the irregular convulsions marking the gradual pre-modern movement toward the city-state. Manuel thinks like a topologist of an arithmetic presenting the historical transformations, occurring loosely as invariant *ricorsi.* The cyclical model builds on an invariant, as a circle cannot alter or vary from its geometric form, without ceasing to be a circle. The danger as well as promise of such a model is obvious; we may think we know our history merely because it is familiar to a present customary way of life, the medieval *habitus.* Yet the hazardous Vichian approach liberates as it limits our wishes, for it accepts that our sense of the past is persistently attacked by failures of human memory, for who can be sure about what precisely different eras saw happening in their own times, let alone at other times? About the past we simply know very little, by comparison to Leopold von Ranke's linear positivist belief that we must search for "what happened," as if we could ever *know it,* and *The New Science* recognizes this deficit and attempts to lessen its effects by adducing a cyclical account of repeated human encounters with returning yet necessarily primordial fear and danger, as if history had perforce to be uncanny. True, the Vichian method gives us only an abstracted picture of the forces of cultural change, but at least he searches for constancy in such change, a constancy shaped in circular form, like the hands on a clock returning continuously to the twelve o'clock position, and while that might seem the most obvious of thoughts, its consequences are profound.

More accurately we could say that cycles of time are closer to the rotations of a sphere than those of a circle. As Etienne Klein writes in *Chronos: How Time Shapes Our Universe,* whilst we exist in a universe of Einsteinian four-dimensional space / time whose character and whose magnitudes we can hardly imagine, we humans actually inhabit a smaller middle scale, usually a tedious world of day and night going round and round, so it is only reasonable to think that time itself—whatever we think it is—must also go round and round. That sensation gives us our immediate world, and it may lead us into conceptual error, for time is not an arrow that flies or a lazy river that flows, but is a mysterious mobile dimensionality of existence. We no longer speak with cavalier abandon about the philosophy of time, yet there is an important index in that clocks do tell time, as their *hands* rotate, symmetrically closing and repeating the circle. Picturing a closed loop returning always along an apparently infinite series of identical steps, a motion in principle without beginning or end, the clock face (along with other such rotations) seems to belong to the being (or is it the existence?) of nature—in any case, traditionally the circle is a favorite Symbol of Perfection. This compulsive religious belief was so strong in earlier times (or are these only a kind of recurrent present time?) that even after Johannes Kepler had shown that our planets follow elliptical orbits, Galileo, who corresponded with Kepler, continued to hold that the planets were following *circular* orbits. Admittedly, this choice may only reflect Galileo's cautious political attitudes toward Church doctrine. On the plane of such debate our deepest questions seem destined always to recur, for they relate to human mortality.

Historically speaking, the circles of orbits and their children, the glassy Ptolemaic spheres, could not have remained the complete authoritative story for cosmic design, for science and beliefs keep on developing, as Ralph Waldo Emerson suggested about knowledge in general and about circles in particular, since after the Renaissance they began to lose their ideal iconic charisma. It is typical of Emerson's complexity that he thought simultaneously of the way we humans inhabit a historical world both linear and cyclical, in that regard recalling Vico's vision of cycles. In his oracular essay entitled *Circles* Emerson virtually denies perfect closure to any circle whose ideal shape cannot preserve a puritan isolation from any other circle, but each one intersects like an astral body or cosmic bubble bouncing around everywhere, along multiple dimensions. The flux of Emersonian circles resembles Richards' *interinanimation of words,* as they proliferate unstoppable ambiguities. In the celebrated essay from Emerson's First Series we read:

> The life of man is a self-evolving circle, which, from a ring imperceptibly small, rushes on all sides outwards to new and larger circles, and that without end. The extent to which this generation of circles, wheel without wheel, will go, depends on the force or truth of the individual soul. For it is the inert effort of each thought, having formed itself into a circular wave of circumstance—as for instance an empire, rules of an art, a local usage, a religious rite—to heap itself on that ridge and to solidify and hem in the life. But if the soul is quick and strong it bursts over that boundary on all sides and expands another orbit on the great deep, which also runs up into a high wave, with attempt again to stop and to bind. But the heart refuses to be imprisoned; in its first and narrowest pulses, it already tends outward with a vast force, and to immense and innumerable expansions.

Exalted and excited this rhetoric may seem, but its circles give a paradox-laden, self-critical description of a person's mind looking beyond its own immediate confines to some wider, different scene. Emerson is actually talking about his own *Essays* here and about the circulatory shaping and reshaping of his own superabundant sense of changing relations between words and their points of reference. In another less enthusiastic manner, rather more reminiscent of Thoreau the professional surveyor, Emerson also puzzled over the desire or the obsessive drive toward immense, unlimited expansion: "The field cannot be well seen from within the field." He understood the need for boundaries, and finally accepted the limits they impose upon perception, although they accept our narrow scope of human awareness. We think beyond what we can experience and live through.

Our Vichian cyclical principle must therefore reflect the following order: *On all scales, large and small, the shape of history that counts most for humans is the physical periodicity bounded by the diurnal and seasonal availability of solar energy. After that all linearities and straightened variants of the cycle, including very large or long-term periodicities, may even be incidental to life, although when sufficiently extended they may beguile us with empty promises or futures of limitless progress.*

My insistence on the Vichian cyclical principle can only therefore be that it has a kind of *natural* priority, a primacy in biophysics, owing especially to thermodynamic biological facts. On this view the environment still grounds life as its encircling surround, such that when its boundary is envisioned in historical memory or felt as an ongoing present experience, our most basic conditions of life (we sleep well or badly, for example) conflict with apparently opposed linear historical ambitions which, if we try to supplant life robotically, will raise questions not only about the movement of being we call life, but about all the forms of value any particular life may be thought to exhibit. Kubrick's *2001,* like his other films, is much concerned with sleep and sleep deprivation, because sleep is where we dream and refresh our powers and is the place (or is it space?) from which we awake.

To speak of night and day, sleeping and waking, dreaming and daydreaming, is to propose the importance of human scale as distinct from all cosmic grandeurs. The issue will be always to think of shapes in history and the *vita activa* as metamorphic and also to think of all the different ways that life exfoliates. The circle in any case proposes that we think about shapes in their most palpable reduction to simplicity, especially that of the ideal two-dimensional sphere, for indeed the circle then becomes the progenitor of our rotating earthly sphere, which in turn is the home of living matter. Certainly one can imagine an historian designing a cyclical model, by fixing upon its conceptual character, and it is above all the Vichian emphasis upon design and custom (the *habitus)* that counts for the Earth's inhabitants, for whom time alone gives meaning to space. In dealing with the time of diurnal biophysical rhythms, as the reader will see in my later comments, there is almost no way to avoid the cyclical, which I am calling "spherical," although such is not the way astrophysicists analyze the morphology of cosmic time and space, nor certainly what historical time resembled during the nineteenth century, after Malthus and Darwin and also the onset of a relentless industrial *Idea of Progress*—J. B. Bury's ominous theme in a famous book, not to mention Spengler's *Decline of the West* and its portrait of Faustian Man.

The idea that history repeats itself is an old one, but Vico asks what such repetitions would amount to, in the long run. What *is* such repetition? The *New Science* probes the cultural consequences of cyclicality itself, connecting the cycles to life on many levels, for as Isaiah Berlin, Leon Pompa, Donald Verene, and numerous Italian scholars have shown in various ways, Vico is no innocently dreaming mythmaker, when effectively he invents the social science of cultural comparisons. Calling him one of "the boldest innovators in human thought," Berlin, in *The Power of Ideas,* says this: "Vico's achievements are astonishing. He put forward audacious and important ideas about the nature of man and of human society; he attacked current notions of the nature of knowledge, of which he revealed, or at least identified, a central, hitherto unformulated variety; he virtually invented the idea of culture; his theory of mathematics had to wait until our own century to be recognized as revolutionary; he anticipated the aesthetics of romantics and historicists, and almost transformed the subject; he virtually invented comparative anthropology and the social sciences that this entailed; his notions of language, myth, law, symbolism, and the relation of social to cultural evolution, embodied insights of genius; he first drew that celebrated distinction between the natural sciences and humane studies that has remained a crucial issue ever since. Yet . . . he has remained outside the central tradition" (2002, p. 53). Vico was in fact one of the first modern historicists, and his attempt to ask questions about the driving forces of progress and recurrence, a question answered by his emphasis on creative imagination, gives him a special place in the history of thought. Effectively he is asking what is the shape of change, which refers us directly to both the arts and the sciences and above all to cultural transformations: thus he takes seriously our common phrase, *the shape of things to come.* What interests him is the quality of existence, rather than the quantity of its elements; such quality admits of endless secondary events, many of them horrendous, some of them salubrious. Maurice Merleau-Ponty described this ambiguous situation as well as anyone, notably in his essay on "The Metaphysical in Man." In the first place an event in history "will always be personal to some extent because it has no basis other than the probable" (1964, p. 92). For the historian there is no easy path to seizing the probable course of past events, and as Merleau-Ponty continues, "It is then that the task of history appears in all its difficulty: we must reawaken the past, reinstate it in the present, recreate the atmosphere of the age as its contemporaries lived it, without imposing any of our categories upon it, and, once that has been done, go on to determine whether its contemporaries were mystified and who, of their number or ours, best understood the truth of that time. . . . We will arrive at the universal not by abandoning our particularity but by turning it into a way of reaching others, by virtue of that mysterious affinity which makes situations mutually understandable" (ibid., p. 92). Such is the intensity of the historical understanding implied by the metaphysical in man, and such was Vico's hope for his own *New Science,* despite its necessary schematism of narrative, for he intended his reader to see that human history is a quality of social "affinity," as Merleau-Ponty suggests.

Topological place and historical cycle are here analogically connected in the way they *both* display permanence in change, the persistence of a recurrent pattern in the midst of reshaped morphology. The question of a larger earthly destiny prevails, but in what direction, with what side effects, whether we speak of the anthropocene or in other terms? We humans share something with cosmologists; we too have a universe to contend with; but it is our own, which we have made, and will perhaps unmake, since we humans alone create its description and its theory, such that Vico remains a guide, because he follows philosophic tradition and modern common language when naming this creation, this making, a poetic activity or *poesis* in ancient Greek. His theory of historical cycles is in effect a topological myth of human development and recycling, and like a theory of organism it points to an idea of what can only be called *the ordering of a complex human emergent.*

If we seek the measure of this emergent pattern, we must evade the paradox of the heap, and the only way to do that will be to shrink the historical accumulation of uninterpretable anecdotes, and the only way to do that would be to imagine history as periodic. The cycles shrink and concentrate history into repeated cultural patterns, as opposed to what Frank Manuel discovered in the endless string of interactions between city-states and other, usually tribal, social combinations. The periodicity is at best a rough approximation, admittedly, the beginnings of a science, and necessarily one appeals to powers of imagination, for whose approximating quest and blurred edges I make no apology, although "the mind imagining" is its own defense and we are free to shrink history, so that we can escape its linear plethora.

Vico shows that we must think even at the cost of wild etymons and patently mythic metalanguage. If we ask how legendary historical writing may be, we are asking how much symmetry between letter and spirit we should or could expect. How raw is the raw data of history, and how long or with what linguistic spices shall we prepare it? By speaking of the shapes and the forming of events, cyclical or linear in sequence, what in general are we saying? Anyone who has conducted psychophysical experiments will tell you that its individual subject's "judgments" are hard to judge. We confront a tautology, where an idea is an idea, a form is a form, a shape is a shape, an estimate is only a reasonable guess—which seems not very encouraging.

Yet in the world we actually inhabit, this apparently circular definition may be exactly what we want, which is not to say that such commonsense local knowledge or its mapping is all there is. The present reading of Vico sees in his *New Science* a rare attempt to fuse the spatial and the temporal, whereby the shrinkage of history—as I am calling the cyclical shrinkage of linear history—derives its immediate force from a cosmic localization of our planet, as a biosphere, a spherical shape which permits and supports the life process. Linear history is not without power and reason, but it does resemble the number line, which can picture endless accumulation—more is not enough—without limits. By contrast, if we shrink linear history, we also shrink a "thick" local knowledge instead of expanding it ad infinitum into the more abstract global range—thus developing a contrast explored by Clifford Geertz, which he called *symbolic anthropology.* In an age of extravagant cosmological discoveries, such as our present period, this will perhaps appear a crude and naïve reduction, but it is not so, for without day and night, spring and fall, we would not be here to read books or have conversations, nor could ants build their heaps or certain beetles in South Africa build the trapdoors to their rolling spherical homes. Day and night produce our mid-scaled macroscopic time, while to the east, west, north, and south lies the mid-scaled space or placement where significant life happens.

The Vichian shrinkage of history does not necessarily suggest that one abandons the larger cosmic scales, when conversely thinking of their reduction in size and time. In his elegant textbook introducing the basic manipulations of topology, *The Shape of Space,* Jeffrey Weeks trains us to think of space-like objects as ideal and empty of material substance, but nevertheless distinctly capable of being deformed or reshaped, as if they were real things in what is loosely called (by non-philosophers) "the real world." At times it seems almost as if this real world is only "real" because it has too much shape, like the galaxies! At the other end of the scale, inquiry must look beyond such first impressions, for although I cannot experience an electron or a virus by gazing more intently, electrons and viruses are nonetheless real enough in a scientific sense, if only in a material, instrumental, optically projected and not psychological fashion, having little to do with experience or what is usually called common sense, and yet, as I have said more than once, the seventeenth-century English visionary and protoscientist, Sir Thomas Browne, wrote in his *Religio Medici* that we live "in divided and distinguished worlds," which inspired another of Browne's books, his colorful pre-Vichian magnum opus entitled the *Pseudodoxia Epidemica, or, The Vulgar Errors*—many of which were his, as he well knew.

Assuming Browne's mysterious ontological *division,* it is primarily the more local world and its flawed common knowledge that interested him as a scene of what he called "vulgar errors," and many of those might stem from mystical views of the astral universe, reflecting the early beginnings of full-blown Newtonian science—the science of action and reaction. Beyond the shadows of Plato's ideal forms, human mortality enforces a constantly changing contour of each distinct example of life. In principle, if one shape (a camel) might generate another into which it is transformed (a dromedary), does this not point to an unchanging original essence or "soul?" A historical argument could be made, that only when topology entered the larger modern picture of our sphere, could its mechanisms of very slow change be understood. To Victorians the later Darwinian picture of change seemed to merge quality and quantity in a dangerous religious fashion, and again we are reminded that topology encodes quality initially, and only then quantity, so that in biophysical terms our cycles of progeneration remain the founding topological fact, situating our thoughts in a mortal context, where imagination always asks what it means to shape our ends and purposes over long or even very short periods of time, as Mikhail Bakhtin's *chronotopes* imply, where time and space intersect on a human scale of cultural interplay between changing stylistic periods or epochs, thus evolving their *chronotopology.*

Meanwhile, the aim of shrinking history to cycles remains twofold: it fits the primary thermodynamic fact linked to the rotations of the Earth, and it opens our analysis to the correct level of cultural interpretation, which must be correlated with our world and not with such distant events that only the Hubble Telescope can record them. In short, by thinking of time and space as the medium of our waking and sleeping, we get close to the idea that we are somehow *placed* in our lives, as historians and poets since time immemorial have shown. We are born and we die in fact, but certainly not as mathematical functions on the purifying abstract plane of counting. Unless, as some religions hold, no individual life counts for much.

VI

"The Round Earth's Imagined Corners"

With magnificent panache John Donne began one of his *Holy Sonnets,* by avowing his skeptical fascination with early modern maps and museum pieces and grandiose ceiling frescos. Four sublime lines begin his chant:

> At the round earth's imagined corners, blow
> Your trumpets, angels, and arise, arise
> From death, you numberless infinities
> Of souls, and to your scattered bodies go . . .

Though its theme is Christian repentance, Donne's is also a hymn to the paradoxes of projecting the shape of the sphere.

One contrast here is a fine disparity. Donne's combination of rectangles at the artful corners and the scientific roundness of Earth present a perfect topological exemplum. We can transform each shape into the other, for the two contrasted shapes merge through a continuous invariance, as with Euler's Polyhedron Theorem, even though visually and metaphorically one is surprised that cube and sphere share a common essence. The two shapes are Platonic solids, in effect. Everything that follows and everything that concerns me regarding a critical ecology flows essentially from similar situations: the Earth may be only an approximate sphere, but its shape is spherical nonetheless and many sections of its surface are locally flat. Its life, motions, and geochemical character depend in a unique way upon this geometry of site, and in an introductory fashion Donne is tracing the immense grandeur, even the mystery, as the Sun illuminates our spherical scene. The distant picture of Earth we take to be clear and distinct when photographed from a spaceship or satellite, may at times appear muffled by our atmosphere, but the shapes of change in ecological experience derive finally from the precise way the form of our sphere makes our lives assume an experienced analogue to certain determining physical facts of our being alive on this planet.

Here a word on approximation may help: even the spherical shape of our planet is not so easy to analyze, and the round earth's shape has everything to do with the vast complexity of life and its changing conditions, which are at once both simple and complex. The Earth has been known to be a sphere since remote Antiquity, a fact that permitted Eratosthenes of Alexandria to calculate the earth's circumference in the mid-third century BCE. Among sailors this spherical picture of the planet became common knowledge and a common experience was of ships disappearing "hull down" over the horizon. Hunters climbing high hills knew intuitively that thereby they would see farther off. With no knowledge of perspective, they knew instinctively that down country the scene would differ, and one might walk toward that different place. Their intuitions about earthly shapes and changes (or invariance) of shapes were of course correct, but only intuitions, and why the ornamental corners in Donne's picture place the viewer (the reader) in a seemingly familiar space, but every point on a spherical surface is equidistant from its center, unlike points on a flat surface, and the result is what art historians call baroque, a torsion. Landmarks then may disorient our sight.

Modern ecology now replaces the stark regularity of "the round earth." While geometrically the sphere is an abstract mathematical object, for the inhabitants of Earth it is physically layered into the atmosphere, the biosphere, the hydrosphere, and the lithosphere—a fundamental fourfold for our existence, giving us the air we breathe, the environment we live off, the water we drink and all aquatic species inhabit, the stony ground on which we stand. The biosphere is plural, as Dorion Sagan's *Biospheres* has described our situation—a multilayered flux of protections and "nourishments" of living as well as abiotic entities, of stones as much as trees. The layering is strong but subtle, and each separate sphere merges with the others in a constant interactive flux. Living creatures inhabit all four domains, but my human focus inevitably is the second layer, the biosphere proper, and yet in the background to my essay looms the thought that humans are its most troubled, if most advanced, species of animal.

Pious commentators might say that humans should be above their turmoil, and we deserve a more delicate treatment than some game of abstract geometry. Yet piety makes problems for factual knowledge, and is more at home with rituals, icons, cults, and authority. No doubt these ultimately religious attitudes have something to do with idolatry, and yet science itself may become a fetish, since more profoundly we may need to accept that deep down we are afraid, if only because, as John Gribbin has described our basic condition: we are "alone in the universe," the question that so long ago Donne's *Holy Sonnet* addressed.

Dividing the Sphere

Making distinctions can occur in any discipline, not least poetry employing metaphors and their disparities. The mind of metaphor, Vico might well have said, inspires his notion of cultural *poesis,* and nowhere is this more the case than in the life sciences, beginning with anatomy. The ancient Greek word *daio* means *to divide*—it provides the root meaning, *daimon,* virtually the life force—a dividing power I have elsewhere shown active in allegory. Creative when daemonic, this calls for an art necessarily involving isolated effort.

Although we think of ourselves and other beings as having real or, as we say, material bodies, our world cannot be fully imagined and controlled unless we adopt ideas leading in the opposite direction. Although we have long since ceased talking about the alchemical aspects of health and wholeness, medical science still imagines the body as cosmos, even when we use the most advanced devices of radiology and ultrasound. In medicine humans are acutely aware that thought may be a kind of action. Usefully, in his final book, *An Essay on Man* (1944), Ernst Cassirer drew a distinction between the symbolic space of culture, the "space of action" where notably the animals often surpass human abilities, and "abstract space" where, analyzing geometrically, we think in conceptual or ideal terms of the material world lying all around us. A triple picture—symbolic, activated, and abstracted—allows for parallels between intellect and our knowing Earth's conceptual edges, such as latitude and longitude or real edges like the seashore. We necessarily must accept a prime element in this understanding, namely the permanent yet shifting shape of the planet's biosphere. Topology in this context is the imaginative as well as scientific key to understanding the layered biosphere where and how humans actually live and may hope to live. The Earth is a kind of body in the most traditional sense, an idea made all too real when we study the visual arts, let's say David Maisel's aerial photographs of eroded and biodegraded landscape, images which are works of art, derived from seeing how destructive humans actually manage to be; in those remarkable, skilled images our planet looks as if it were bleeding to death.

The integral form of our bodies is like the sphere of earth, and if not torn, it will survive, alive, up to a point—at least from wounds and internal injuries to the organic system. Climate change mimics larger wounds, as one learns from Paul Edwards' treatise, *A Vast Machine,* where he summarizes a very large body of research into climate and its measurement. When research builds such rich bodies of information, it is not easy to distinguish between instruments and bodies, much as when using new diagnostic devices, well beyond X-rays, modern medicine also blurs the edges between the organic and the mechanistic, and this blur in fact saves *lives.* What such statements mean is not yet clear, but the usage is itself a fact.

It seems no accident that a designer and maker of his own instruments, James Lovelock, has pioneered a crossover between life and life's self-observation. Lovelock has argued for a broadly holistic view of global process. Going by the familiar name of *Gaia,* following a suggestion from William Golding, the Nobel Prize winning author of *The Lord of the Flies, Pincher Martin, The Inheritors,* and other novels of the extreme condition. As we shall soon see, from comments by Dorion Sagan, Lovelock never intended to be mythical or mysterious with his adoption of Golding's literary and also literate suggestion. By insisting on the feedback loops in complex adaptive systems, as we shall have occasion to note, Lovelock has argued that Earth in some sense can only be imagined, and properly studied, when the whole Earth is seen as greater than the sum of its parts, although not more complex than its controlling, emergent cybernetic loops. A professional geochemist, inventor of his own instruments, a general thinker on the highest level, Lovelock is utterly unlike the Californian New Age mythmaker his opponents have at times called him. Lovelock is a great example, because, technically accomplished though he is, he does not exploit the imaging of Earth as a massive data-mining operation, but he has sought its broadest theory, given its manifestly organic life process, when by organic we understand a cohering life-system on a very large scale.

To sense Earth this way, we have to think topologically, for that will reveal systems of transformative connectedness, in the midst of apparently radical change; otherwise we will lose sight of the complex interweavings of biospheric physics, for it is the wholeness and completeness of the biofeedback machine as an embodiment, that enforces the facts of geophysics and geochemistry. Recall that imagining something or some state of affairs is not the same thing as measuring it or boxing up the sum of its parts, but must rather engage with a unifying esemplastic purpose. On the other hand, developments of topology do permit measuring biospheric changes across the outlines of given shapes, which we call "climate change," and nowadays the algorithms of computer technology are able to implement such advanced programs of measurement. Current topology, which since Poincaré has used an algebra of point-analysis, easily describes edges. We can now employ a measuring topology, and we can imagine the outcome of a certain measurement of shapes, which we can also "see," as they move, stand still, or are crossed by dividing lines. Yet the modern scientist is committed to measuring everything, cutting nature's products and processes into countable pieces. In connection with the fact / value dichotomy, Hilary Putnam has suggested more than once that science and its measurements have long been locked in dubious battle with valuation, the latter identified with a sort of combinatorial fancy, more like pictures in a dream than any so-called hard fact. Yet there remains a biotic need for the *eidolon,* to use Walt Whitman's word for the living form, and that surely implies value.

For living creatures to survive they absolutely require the dynamic play of sense, as our senses combine to identify what is called Life in other creatures. On the artistic side, music yields good examples: Igor Stravinsky was famous for wishing to know how many minutes a composition was expected to last in performance: perceived reality is complex, endless, interpretable, and finally shapeless to all but the gifted eye and ear. As a student of musical manner, Stravinsky is interesting, most especially (like Eric Satie!) in contrast with Wagner. Consider the revolution represented by Wagner's unbounded *Gesamtkunstwerke* excesses, when he practiced "the music of the future," which Rossini thought ridiculously pretentious; opera as cosmos lacked wit and economy, which Wagner's extreme chromatic methods had relegated to the dustbins named Haydn, Mozart, and

Schubert in favor of the *liebestod* in *Tristan und Isolde.* (Flagstad was wont to sing long stretches of *Tristan,* facing away from the audience—actually a clever reading of the opera.) Pausing on Wagner for a moment, I wonder if edge did not also help him to measure quantity, restraining it, although Wagnerian chromaticism can expand musical quantity to such great lengths, as if to say, why should this scene ever end? Maybe that is a new note of nineteenth-century quality, when it comes to the universe. Perhaps with *Das Rheingold* or *Götterdämmerung* the distinction between quantity and quality collapsed, but inwardly we may wonder if somehow the difference remains.

Incisive Terms

When it comes to the language of incisive distinctions and divisions our lexicon is almost too rich. "Cut" is what William Empson called a "complex word." We speak of distinct or newly found significance as a cutting edge; there are cutting boards and cutting remarks and "the unkindest cut of all"; assassins cut off their victims' heads; armies are cut off at the bridge; someone does or does not make the cut; being chosen is "making the cut"; there is the cut of a suit or "the cut of his jib" or a clear cut in forestry or just a plain haircut; cutting a deck of cards or cutting a class are not as different as one at first might notice. The list goes on and on. "The unkindest cut of all" is only the beginning. A keen look at the *Oxford English Dictionary* or the original Roget *Thesaurus,* with its late nineteenth-century ordering categories (but without definitions), shows that edge-terms are linked to controlled movement, and hence to strength, energy, ridges, shores, boundaries and borders, dimensionality, directions, trends, deviations, flexions, success, sharpness and roughness as its antonym, desire and feeling, nervousness, interjacency, progression as in edging forward, obliquity in beveling, ornamenting objects around their borders, and so on. Lacking a concept of edge, there is no way to think of closures and openings or even blurred thresholds, nor can we map geographic boundaries without such an idea in mind, or without a means of graphing or tracing it as a central component of the original Königsberg insights into positions and sequence.

Once again we launch forth into the sea of paradox and metaphor, since Earth conceived as an ideal sphere has no edges, while its face as an inhabited, constructed, or even wild natural scene consists of nothing but edges. To grasp the meaning of such a paradoxical case is the task of imagination, a topological matter rather than a question of measurement, for as Euler had shown, his new geometry of *situs* was designed to avoid all measurement, focusing instead on the shapes of things and situations. His new emphasis leads mathematics into judgments of quality, perceptions an artist might instinctively treat, and these perceptions tend to resist logical treatment and aesthetic awareness operating simultaneously. On the other hand mathematicians have frequently insisted on the ideal aesthetic aspect of definitive proofs! Particularly in dealing with the ideal versus actual character of our planet as a sphere, we encounter a strong feel of virtual contradiction, since the surface of our planet is far from ideal perfection and regularity, being too full of massive sudden rises and falls, ocean canyons and terrestrial mountain ranges.

Geological time is slow and terrestrial shifts along fault lines do serve an edge-function, providing a dangerous belief in a stable crust. Topology suggests that the one-dimensionality of the edge permits it to define objects bordering on visions of "place," which is to say more generally, edge creates a language of place. Within civilizations where society inscribes boundaries, there can be no such thing as entirely unnameable space or spaces where emotions are denied, as inhabitants of deserts will testify. Boundaries are basically disturbing topics, if we may pun on the subject. Always, the inscription of edge is as sharp, and often disturbingly so, as Euler implied when he named his edges by his Latin term *acies.* Boundaries may be crossed, but they also divide, and they have their preferred cutting icons.

For that reason Sir Walter Scott opened his 1825 novel, *The Talisman,* with two warriors—the Saracen and the Crusader—displaying their fearsome weapons, each capable of cutting the enemies to pieces. But how different their weapons were to handle! The Saracen carried an elegant scimitar, which could slice a cushion of down as easily as cut a man's head off; the Crusader's massive broadsword was capable of splitting a stone—the one (we might say) abstract and mathematical, the other material and ponderous. At one stroke Scott captured a romantic version of terrorism, understandable as the arming of two opposed visions. The blade runners of history carve the shapes of history, and Scott the historian had studied wars of religion. He understood their savage ironies cloaked in *true belief.* Surely with him we were meant to sense the shifting momentary edges of existence, and when we trim our thoughts to the wars of love, we find the same study. Shakespeare's *Sonnet 116* avers that "Love bears it out even to the edge of doom," containing all life itself. This edge is Prince Hamlet's "bourn from which no traveler returns," whose crossing resembles nothing so much as the ancient Roman walls around Rome, which they called the *mundus,* the world a great city was intended to contain. Inside the wall one was safe and secure; one was much less secure beyond those city gates protected by the god Janus, whose statue faced out of the city and also, symmetrically, into the city. January was the annual edge, a month facing back into the past and forward into the future, the known and familiar and, outside, the unknown, while the critical gate at the boundary is always uncanny.

None of these terms and meanings, ramified in many directions, could begin to occur had humans not devised the definition of security by understanding *the edge between,* which surfaces most seriously in the primordial rites and meanings and purposes of burial custom—virtually unique to humans, as Vico said, in *The New Science.* Mary Douglas's *Purity and Danger* showed in recent times that such customs mark the ominous boundary between the pure and the impure, in extreme cases between the sacred and the profane. Human burial rituals incise an edge between what is safe and what is dangerous, between the living and the dead. Edge also cuts away from unmeasured spaces a line of junction whereby we may sense whatever is local or nearby, a purpose essential to forces acting across any widened field, for example an electromagnetic force or the dilated *ecotones* discerned by ecological science.

Finally, the edge has its basis in the topology of what is most important of all imaginative achievements, namely the idea that we must think beyond our limited bodies by troping and bending the fixed image that occurs to our first perceptions. Topology as the science of turning, twisting, bending, and stretching serves a function, the turning of the trope and the tropic, as we know from our heliotropes and all the phototropic plants—all of these involve a turning to the light, and almost always sleeping in the dark. Theologically, in a conversion, ideas of spiritual enlightenment hinge upon the exact moment of turning from one faith to another, as with c*onversion,* and there also we discern a residual permanence in change, coloring the new beliefs with old beliefs. As Hannah Arendt and other thinkers have shown, all powerful revolutions are fed by carefully controlled features of their past, dilating their changes with deeply hidden inheritance. The dividing line is never a distinct geometric cut.

Finally, in nature there is no way for us to avoid the determining role of edge in all the processes of nature's interactions, so long as we allow approximations, not even in the waveforms of quantum mechanics; we know the physical boundaries statistically, though with great precision. In all such cases the observed edge (say, between one cell and another) will define a dividing dynamism, some occasion or event where energy is expended and where forces exert possible opposites and disparity actions. This dynamical aspect of edge is what makes the spherical situation so important, since edges occurring on the sphere, which theoretically prohibits such lines of demarcation, edges occurring where geometry says there is no such possibility—*the sphere has no edges*—reminds us that nature does not always obey axiomatic rules, though such rules may help to illuminate nature's course. Clearly physics and dynamic factors give the lie to any ideally spherical prohibition of edge. But then, in this situation, why do we still speak of edges? Perhaps it is only a matter of mind, to speak paradoxically. The idea of critical discrimination is here to be given its root sense, that of conceptual lines that define boundaries of all sorts, abstractly at times, at other times, concretely, as with walls. In these terms the chief reason for focusing on edges is that without them we could not think where and what we are, probably could not think at all, since we live on a spherical planet which in principle obscures its natural edges, or else blocks thinking about them; we would have no method of making judicious discriminations between subtly different states of affairs, or, to put it bluntly, fairly adjudicating quarrels over disputed territories and spaces. We could only deal with indecision. Power hunger always prompts a desire to recut borders, but without the basic notion of incision we would have no idea of the life process, nor of its beating heart.

Topology and the Circulation of the Blood

Like our planet Earth the human body is almost a "deformed" sphere, yet spherical in some topological respects. Not only so, but the heart and all its mysteries of anatomical connection played an early role in conceiving the theory of knots and loops. That such was the case for early modern science was not solely physical in nature, however. A theory of heartbeat in fact drives a *vitalist* political vision, as John Rogers shows in *The Matter of Revolution,* a study of William Harvey, Andrew Marvell, John Milton, and other luminaries, noting their affinity with a special type of revolution, namely the circulatory model whereby the heart pumps blood through the arteries and veins. Rogers is particularly interested in the monarchical aspect of the heart, as if, sharing the life force, it were a king of circulation, whose idea comes to influence political thought during the later seventeenth century. The "body politic" then provides a strong analogy between medicine and (after many years) an ultimately liberal sense that the heart of the political body is a stable, driving, pumping organ of political coherence—feeding and refreshing its inward environment, and thereby affording government a circulation of vitalizing power. Things may get to be exceedingly complicated when we consider the flow of living, biotic substance, with its almost infinite mesh of living cells and all their interactions, but "Spherical" once again here means a closed but not necessarily simple circular loop, loosely resembling the system of the circulation of the blood, first described by William Harvey in 1628.

Harvey was famous for his remarkable vivisectionist observations, performed without sophisticated lenses. These observations made on animals, fish, and birds enabled him to analyze the function of the heart's "universal" capacity to pump the blood and lymph, through the complex pathways of veins and arteries. Through its continuous circulation the blood not only visited remote parts of the body, but more importantly the heart pulses moved oxygen throughout the whole biophysical system, animating it, one might say, while momentarily also reanimating the "tired blood" through a secondary process of reoxygenation. The air we breathe is almost the main beneficiary of Harvey's looping model, which to this day remains a process of life-giving circulation, in short, a topological circuit enabling life itself, where overall a topology of the biosphere emerges.

On a global scale, but no less involved in revolutions of the breath of life, a much later modern development for the networked, even politicized character of ecosystems occurs with the work of Vladimir Vernadsky, as described in his book, *The Biosphere.* The analogy to the heart suggests a vital (if not "vitalist") context for discussing the topology of the terrestrial sphere, which recently was demonstrated by James Lovelock and the late Lynn Margulis. In *Dazzle Gradually: Reflections on the Nature of Nature,* Margulis and her collaborator Dorion Sagan show how Lovelock's Gaia theory works, and they show how the earth's atmosphere virtually manages (or at times mismanages) a chemical disequilibrium of many orders of magnitude, such that life-related gases are "produced in prodigious quantities and turned over rapidly." This rapid seemingly tautological turnover is one way of understanding a large-scale circulatory system, here a global atmospheric flow, and as Lovelock carefully demonstrated, it can thus be measured. Finally what applies to atmospheric conditions applies in parallel heat-exchange fashion to the oceans and lakes (the hydrosphere) and even to the upper layers of our stonier "ground" (the lithosphere), where in the soil of the fields and valleys there is a truly infinite vastness of submicroscopic organic biotic material, all together feeding into the larger circulation comprising a coherent life-system. By analogy, we live in a world of branching arteries and veins, whose heart is the light of the Sun, and not for nothing have humans identified "the breath of life." Nor is there anything sentimental about this analogy, as a man dying of emphysema will tell you, if he can still speak. Or more modestly, a hill climber "short of breath." There is perhaps no need to continue with the idea that blood breathes, an ever turning and burning system.

To admit the circle of argument involved, we need to escape from narrowing provincial habits of mind, for as usual we meet a disparity we must deal with. The circling flow of the blood, an almost universal model for the processes of life, is still interrupted by a breaking of the circle, for the circulation cannot protect itself from outside influences—from climate changes, for example. Like the diurnal and seasonal circling that our planetary rotations ensure, our looping ecological circles are not permanent adamantine walls providing an eternal Roman *mundus.* There is no Great Wall that will block unwanted changes, so that circulations enforce emergent changes within the evolved environment.

For a human body the "spherical" means after birth a three-dimensional, opening and closing, extremely complex spiraling loop, an embodiment of changing life in actual macroscopic fact, whose rivers and streams of pulsating circulation William Harvey had discovered. There is nothing simple about this circuitry, indeed it may be the most complex array of processes we know, and yet it establishes the primacy of cycles, none of them ideal geodesic pathways, but all of them returning periodically. Every time we breathe, our bodies enact this regularly recurring permanence in change—an invariant life process whose pulse defines human health.

Biospheric Circulation

A loose analogy between atmospheric circulation and the cardiac process—Harvey's circulation of the blood, in this case—creates a vital context for a larger order, when applied to the terrestrial sphere, and this model of life-process is inherently a kind of circularity. It is also inherently thermodynamic, as Vladimir Vernadsky recognized, when he wrote his epoch-making book, *The Biosphere.* Remarkably, James Lovelock was not familiar with the Vernadsky volume when he devised his similar *Gaia hypothesis,* yet for years he had studied the biochemistry of the whole sphere, inventing numerous instruments for measuring the data. In the MIT Press book, *Scientists Debate Gaia: The Next Century,* a variety of critical positions are reflected, all of them giving credence to the belief that Lovelock's interactive geochemical approach (the Greek name is not important!) is a going theoretical and practical concern.

To determine the parameters of the geochemistry of our planet is an immense, if not a theoretically impossible, task. There is general agreement that we need a paradigmatic shift of ideas toward the extreme complexity of our planetary feedback loops and their prodigious power over climate and all similarly global phenomena. The net effect of climatic disequilibrium is large-scale circulatory weather system, a life-giving global atmospheric flow, almost as if Nature had been reading Heraclitus the Pre-Socratic.

Finally what applies to atmospheric conditions applies in parallel heat-exchange fashion to the oceans and lakes (the hydrosphere) and even to the upper layers of our stonier "ground" (the lithosphere), where Nature amounts to an immeasurably various system of circulating exchanges between biology, subsurface and local climate.

Recent climate changes have shown that Nature owns us, every bit as much as the reverse—we just pretend otherwise. A topological analogy here suggests its abstract relevance to the earthly condition. One consequence of the planet's spherical shape cannot fail in the long run to promote a model of the global unity not of races, but of life processes, whose ecology is a condition as a global climate change, at all biospheric levels from underground to the stratosphere, and to some degree beyond. The total biotic system, as a thermodynamic system, is indirectly linked to Grigori Perelman's recent solution to the Poincaré Conjecture, although one sees little enough interest in the initial heat-flow aspect of the Ricci Flow equations. Poincaré had conjectured a very difficult topological transformation of a sphere into a dodecahedron. After years of mathematicians struggling to solve the conjecture, Perelman developed the brilliant geometrical insights of William Thurston and Richard Hamilton, by following their emphasis on Ricci Flow. Perelman and his predecessors thus used the Ricci equations to solve a purely topological puzzle that had resisted solution for one hundred years.

My interest in them, however, is different: these special equations were invented in the late eighteenth century by Grigorio Ricci-Curbastro in order to express a problem in physics—the movement of heat from one body over to another. More deeply these equations expressed aspects of curved Riemannian space, where space, however empty and abstracted we may suppose it, is still curved in form. Although flow from one state to another is their basic formal principle, in physics the Ricci Flow equations imply a relation to the fundamental process of life, the flow which is a thermodynamic play of heat exchanges, that is, of energy relations. Ricci-Curbastro was also specifically concerned with the movement of heat flows over curved spaces. In this context the globe of Earth demands a thermodynamic analysis. Despite and beyond all technical complexities, the result of Ricci's approach via flow patterns, this topological coherence is a further example of shape guiding a kind of generalized Aristotelian vision of the mean, but a mean that constantly shifts position, which is inherently a nature-derived concept. Whatever Earth is, for living beings it can only be the result of a thermodynamic, recurrent sequencing of events, and in that thermodynamic perspective we can better grasp Gaia and its topological premise that earth's balances depend upon systems of loops, knots, ties, and splices.

The Gaian model is based on remarkably detailed geophysical knowledge. Bitter opposition to the model from differing academic experts will be a familiar experience to any scholars who venture beyond specialist concerns. Altogether there can be little to question about the background to Lovelock's proposals—the dominant fact of present day industrial interference with "natural" balances, as when in America we suck aquifers dry or divert snowmelt resources to promote distant real estate development in desert places. (The interferences themselves are never one-way paths, so at times natural conditions improve, unaccountably, or by our combining restrained consumption with better water management. In either case the human mind is closely involved with exploiting natural resources, in good or bad ways.) Why should we not accept the name Gaia, since it comes from the ancient naturalistic sense of *Ge,* which to the Greeks meant "earth," giving rise to words like "geophysical" or "geography," terms respectable enough, one would have thought! Scientists seem not to object to the idea of *geo-metry,* but its Gaian aspect needs affirming, as related to origins and destructions of living ecosystems. There is nothing New Age about this partly holistic view.

Water—traditionally the "water of life"—plays a key role here. There was an ancient deity of ponds and lakes called *Limnos,* but that myth did not prevent the eminent geophysicist, G. Evelyn Hutchinson, from calling his field "limnology." Some have called Hutchinson the father of modern ecology, for his range (observationally as well as mathematically) was remarkable. This I know from my own experience as an undergraduate, when Hutchinson was teaching undergraduates the art of combinatorial thought. He lifted one's vision far beyond mere naming this field or that field—one was transported to questions about the origins of life. Later, for deeper reasons and concerns and without a marketing plan, Lovelock accepted the name Gaia.

We humans legitimately use personifications when we need to express some odd mixture of thought, consciousness, and passion that marks the "personhood" of all living creatures, from the simplest to the most complex. At the same time both science and our cultural practice justify the Gaian personification, accepting that the Earth is a partially self-regulating *complex adaptive system* of feedback loops, as complexity theorists would say.

Just ask a fisherman if this is not true! The planet, if not what we usually imagine an organism to be, might as well be one, a creature endowed with a bounded life similar to that of an onion or an orangutan, clearly alive in the sense of self-regulation at manifold levels, constantly moving, breathing, and shifting its ground. While the terrestrial globe is a David Bohm "implicate order," a virtual superorganism, its combinatorial feedback functions make it something more like a whole universe of living creatures talking with each other, like cosmic honeybees bringing news from a foreign country.

Combinatorial activity, coherent or incoherent, seems to be the cosmic game. On Earth the life-system recycles its waste and it recombines through a perpetual signaling of macroscopic resources, such as rainfall, to produce something even more important to acknowledge, namely that while the Earth is a massive microbial ecosystem, as the scientists argue, this system in which we participate also involves extremely small organisms. On average, indeed life is microscopic in scale, so that we humans, who are tediously macroscopic in size and scale, will normally be unaware of the hidden forces at work. These interactive biochemical forces in turn provide the networks and links of interactive energy exchange, as if our biological story were a novel by Marcel Proust, reveling in social detail, including the undoing of its own process, since the terrestrial ecosystem depends upon a vast network of shared signals passed back and forth throughout the whole biosocial "body."

Time chronicles these signals passing around and around. Dorion Sagan puts it bluntly, without romantic illusions but with apposite technical detail, when he says that *the Earth acts as if it were a body.* Proust similarly writes by analogy; we are all living in a search for lost time. We should remember that although we may not have read the philosopher Hans Vaihinger, we humans live according to his *Philosophy of As If,* for that is what all action is (though certain realists claim otherwise). That's also what signaling means—"Please do this, as if you understood my message." Of course, that as / if was quite obvious in the days of telegrams. We are never completely sure about the message, or its audience; we are only listeners hearing a lot of static—the *son parasite* described in detail and extensively by Michel Serres.

In the midst of static, human life spans carry the major message. Like the fabled seven ages of man, a Gaian life-picture displays recurrent cycles such as waking and sleeping. As individuals we are all Gaian planets, wanting to resist the Second Law of Thermodynamics, hapless finally though the wish may be. This sounds like poetry and figures of speech, but that is exactly the point—the issues for a global perspective on the planet are always, to some degree *crises of analogy* and of adjustment between like and unlike; and furthermore, they are rarely significant if they are isolated as matters of the large and the small. Global climate change may be vast and gradual, but its explanation requires a tremendous range of observation spanning the large and the small. Mere size of single objects is not very interesting, except to giants, and even there the problem is relative and implicitly biblical.

Lovelock has the best of the argument, which is to ask another question: in *The Revenge of Gaia* he recalls Clerk-Maxwell having trouble in accounting for James Watt's use of a ball-governor to control the speed of a steam engine. "How does it work?" Maxwell wondered, as though he could only guess the answer. The answer lies in a rejection of the standard Cartesian notion of cause and effect sequences. Lovelock continues: "Maxwell's puzzlement was not so surprising. Simple working regulators, the physiological systems in our bodies that regulate our temperature, blood pressure, and chemical composition, and simple models like Daisyworld, are all outside the sharply defined boundary of Cartesian cause-and-effect thinking" (2006, p. 37).

The mind-body split is mathematically or rather geometrically appealing as *l'esprit géometrique,* central to French tradition, but it does not readily meet the actual difficulties of understanding emergent properties such as are exhibited by the complex adaptive systems of life. The failure inheres in the Cartesian split itself, for we can now see that humans *involve* or *infold* our thoughts even when we distance physical objects from our own consciousness, seeing them as outside our minds. As Heisenberg put it, "Natural science does not simply describe and explain nature; it is a part of the interplay between nature and ourselves; it describes nature as exposed to our method of questioning. This was a possibility of which Descartes could not have thought, but it makes the sharp separation between the world and the I [*das Ich]* impossible" (1962, p. 55). In *The Revenge of Gaia* Lovelock discovers the same impossibility, familiar to us from the Uncertainty Principle, as the much older Watt's invention of the mechanical governor showed: "Whenever an engineer like Watt 'closes the loop' linking the parts of his regulator and sets the engine running, there is no linear way to explain its working. The logic becomes circular; more importantly, the whole thing has become more than the sum of its parts. From the collection of elements now in operation a new property, self-regulation, emerges—a property shared by all living things, mechanisms like thermostats, automatic pilots, and the Earth itself" (2006, pp. 36–37). The problem for perspective is that we have always to imagine a larger complexity before turning back to a theoretical center, where all is simple once again. As *The Revenge of Gaia* comments at the end of its second chapter, on the nature of self-regulating loops, "our comprehension of the Earth is no better than an eel's comprehension of the ocean in which it swims" (p. 38). Of course not, since eels swim without what we mental and verbal creatures call "comprehension"—the eels just know. Life and our universe and even simple things like riding a bicycle are, as Lovelock continues, "inexplicable in words. We are only just beginning to tackle these emergent phenomena, and in Gaia they are as difficult as the near magic of the quantum physics of entanglement. But this does not deny their existence."

My own conjectures clearly have a parallel yet poetic side, for they rely on the assumption that for humans the most important emergent is to develop *a sense of the hungering self, given limited time,* which often we call consciousness, particularly when actively sensing, thinking, or perceiving the real-world consequences of our own actions and thoughts. Getting rid of our thoughts, whatever they are, seems impossible, since to be conscious is to experience a feedback loop of experienced reality in which we participate, in which we are often unwitting partners. Beset by uncertainty, our conscious self-awareness unavoidably seeks shaky comfort in the belief that whatever is known is "out there," whereas in fact it must also be "in here," and thus must share in the mental dynamics I am associating with seeing that whatever we know as fact and thereby operate upon is more copious than our narrow expert knowledge, only truly fact when we allow that mind lives both inside and outside, residing in between—finally because the mind must accept the enclosing biosphere sheathing our sphere, such that life may circulate.

The scale of such investigation has not always been sharply focused. When Vernadsky published *The Biosphere* in 1926, it was the first coherent (if frequently schematic) geophysical account of this biotic process and structure. He drew upon his knowledge of geochemistry and on the science of thermodynamics, wherein one could study the flow of heat as it distributed energy (ultimately coming to us from the Sun) directly and indirectly into what Vernadsky called "the living organism of the biosphere" (1997, p. 27). There is a somewhat obscure personal history at work in regard to extended meanings of the basic Vernadskyan concept, the biosphere. In 1922–1923 he was lecturing on geochemistry at the Sorbonne, when he would have almost certainly encountered the work of Pierre Teilhard de Chardin and Edouard Le Roy—distinguished scientists with mystical leanings, we might call them. The spirit of Henri Bergson hovers over the two French thinkers, but altogether, like their Russian visitor, they may be said to believe that the complex web of biospheric interactions amounted to a kind of global mentality, a network of mind. The interconnection of thought between all three of these men would itself make a chapter in the history of theoretical science, but there is no doubt that as a strictly fact-driven geochemist, when Vernadsky spoke of the life process animating the whole earthly biosphere, he increasingly tended toward a wider conjecture—the *noösphere* or mindsphere, which structures a planetary network of biotic processes, including our own interference with Nature. In the present context one could say that, by studiously avoiding magic thinking, we may claim that the biosphere and its mindful double, the *nöosphere,* imagines or models its own cybernetic limits, like an organic and self-aware spherical heat engine.

Living in Spherical Time

Scale and embodiments of scale are always time-dependent, because like the weather everything in nature is always changing, and physics studies that motion. This is true for the smallest particles, and macroscopically the exchanges will be grossly material in kind, as when whole cargoes of food or fuel for burning are shipped overseas to fuel a distant continent. These slow-moving energy transfers may also occur in very fast-moving transfers of information, when encoded words, symbols, and numbers crisscross the planet, as if drawing endless arcs of exchanging e-mails. The global pattern of these electronic transmissions appears static when printed on the page, but in use they represent a vast weather system of circulating knowledge and opinion—the *worldwideweb,* in fact. The Web is an instance of the modern Kantian sublime, a *hyperobject,* as the literary scholar Timothy Morton would call it, a new sort of allegorical matrix. On the evidence so far, when for common assessment the masses of data accumulate too fast, a confusion of numbers with words is likely to occur, and from there another change ensues—a ritualistic fixation on number, as if number alone could enshrine value by its manifest utility in virtually instantaneous calculations of magnitude. Economic inequalities in real-life conditions, such as food and water supply, reflect this; a wealthy woman once said to me in Tucson, Arizona, that she was not worried about the water supply (her back door bordered on a golf course), "so long as we can afford it." It would be easy to show that much of our current worldview is allegorically anaesthetized by large dollar balances and the culture of mass-market selling.

In any case, from the least to the most massive and hence most impressive numbers, the Web's communicative size and speed have the effect of changing the nature of experienced time, from linear to circular to spherical. Distant time zones become present realities, or perhaps we should say, moments of decision as with stock trading. The poet John Clare's personal "knowledge," so real in the nineteenth century, now has no home. A formal issue remains to be explored: can there now normally be a selfhood, a personal existence like Walt Whitman's "I, myself?" Does electronic digital communication slowly annihilate human character? Instant messages and communications certainly block a primitive shaping function of place, and instead promote spatial noise, rather like the town dump, where, as Wallace Stevens once wrote, you were always looking for the definite article—"the the"—*le son parasite.* Recent instrumental advances have reduced the electronic world to a shrinkable sphere, but they have the further effect of collapsing our ideas of time, a collapse psychologically critical, because our world-time seems to be accelerating faster and faster.

Time then, not space, is the scene of gravitational force. When Einstein remarked that time is what clocks measure, he was also saying that space creates gravity. The clock's hours and minutes permit measures of the fourth dimension, which, writing to Henri Bergson, he called "physical time" as distinct from Bergson's "psychological time"—the Bergsonian time-sense so interesting to Proust. Things "going round and round" give us recurrences moving us toward the end of our reckoning the time. For our human macroscopic worldview, however, while we are here, living on the scale of human existence, there is for us a connection between temporal iterations caught by the hourglass, the mechanical or modern atomic clock and the diurnal periods of our lives; all of these periodicities together suggest a strong analogy between our measured time and the rotations of our planet, or to say it differently, *our clock face becomes spherical in form.* As we live our lives, Bergsonian duration *(la durée)* helps to define the shape of consciousness, the *Where* of our thoughts.

My interest here has been to recall that only through Earth's rotation do we know the basic rhythms of our actions and consciousness, reflected in our own powers of rest and movement, these being the twin modes of the daily *circadian* and the seasonally *circannual* rhythms by which living creatures *respond* to light. We abstractly theorize time in the theory of relativity, but our lives are bodily clocked and biophysically formed by alternations of light and dark, so we should not be unduly impressed by irrelevant light-years in the billions of billions! It may not be a good joke, but one might ask, if time is money, whose money is *that?*

On a rather more modest scale, our human time-sense deriving from planetary rotation, which also includes the constantly reappearing Moon, is affected by common use of arbitrary time zones. These zones reach from East to West, but also from North to South, and for that reason alone our full time-sense is spherical rather than circular. Russell Foster and Leon Kreitzman's *Rhythms of Life* (2004) explicates the research into the biological clocks and the rhythms that control daily life for each living thing,when sensory receptors enable a periodic response to changing light conditions. Circadian periodicity is the "rotation" of daily or diurnal waking and sleeping, something we all experience, especially when we have trouble sleeping, but what is most striking about the clockwork of its rhythmic pulse is that it exists on an immense range of size, appearing in the largest mammals down to the smallest bacteria. For example, "The circadian clock in cyanobacteria not only keeps track of circadian time in exponentially dividing cells; it also 'gates' the cell's activities. The clock controls the actual timing of cell division" (Foster and Kreitzman 2004, p. 166). In comparison with common human activities, such as opening our eyes when we awake, the bacteria are displaying an almost unimaginable delicacy of response—but the scale difference is no more important, we might say, than the relatively crude stirring of our bodies when a bright light or loud noise awakens us at night. Both cases, from large to small, involve operations of the living system of circadian sleep-rhythms. The same alternation, but on a larger annual scale, occurs when the seasons change, thus giving rise to what may seem anomalies as we approach the two poles, for instance reading a book without artificial light late into a far northern summer's night.

Such variations of diurnal and seasonal periodicity nonetheless do not prohibit adaptation with the larger patterns and, as Foster and Kreitzman observe, "we all have an individual chronotype or time-signature," which, we have said, should recall Mikhail Bakhtin's culturally defined *chronotopes* (ibid.). We each try to adapt to the cyclic changes within a certain personal choice of time and behavior. Some winter athletes do well in late afternoon, others only at the crack of dawn. The British Empire may have flourished because many of its soldiers grew up under the tutelage of fanatics for cold baths, which saved money, of course, and drove them out to govern hotter lands—who knows? Living creatures at all scales of size and biological organization choose or are forced to fit into their climates, hence may be said to inhabit certain patterned locations on earth, adapting to them, be these as different as tropical forests and arid deserts. Nevertheless, the Earth rotates on two variant planes—the direct twenty-four-hour rotation on its own axis, producing the separate days and separate nights, and the oblique (off-axis) annual rotation around the Sun, producing the Northern and Southern hemispheric seasonal differences, with an Equatorial sameness of non-season occurring between North and South, having "crossed the line," as sailors were once wont to say. The known facts of our rotational planetary movement are comparatively not very complex, but their results are, nor should we blame ancient mythmakers for not yet knowing the science.

Foster and Kreitzman in a subsequent volume have surveyed and explained the data and theories covering this topological area, the seasonal aspect of life. In their book, *Seasons of Life: the Biological Rhythms that enable Living Things to Thrive and Survive* (2009), we learn voluminous details of a larger system of cyclical behavior patterns, such as migration and hibernation. We find that just as single organisms are systems of internal competition and self-organization, so on a broader canvas species as different as grasses, cacti, and wildflowers coexist in a complementary relation of exploiting seasonal changes in amounts of available nourishment and light. In sum, the twin rotations of Earth together make possible an ever changing exchange-system, which is best described as a vastly complex symphony of circulation, where nothing is ever exactly the same as it was in the previous rotation—no day is ever the same, no season is ever the same—and therein lies the secret of life as a variation of tones and harmonies, as if life were an exfoliating commentary on a single mysterious narrative. There is, of course, more to contemplate than the diversity of the biological sphere—the full biospheric life-world in Vernadsky's sense, with all its *dramatis personae*—which is also the heuristic of a Gaian climate network. We humans can hardly avoid the imaginative problem of considering time *in itself,* whatever that means, when we discuss circadian and circannual rhythms of living organisms. At such moments of wonderment our thoughts pass beyond our own small lives, to the seemingly larger aspects of temporality. We are asking what it means to diagram or to graph, and therefore abstract, a living scene in the actual environment, rather like a student of baroque music in the period when Couperin said, "We do not play music as we write it down." The time of performance is not the time of analysis, which leaves us facing an impossible puzzle, only avoidable in math.

In the *Principia Mathematica* Newton stated that "absolute, true and mathematical time, of itself and of its own nature, flows equably without relation to anything external." In the Special Theory of Relativity time is yet again to be understood differently—as a mathematical and then physical fourth dimension: in both cases we are describing what we have, in mathematics, written down. Halfway between the mathematics and ourselves, metaphors for time abound, just like the Newtonian image of equable flow, which suggests a broad river. When collapsed into an idea of an eternal unchanging extent, time could be identified with oceanic breadth and stillness, not an experience of sailors such as Captain Bligh. Nor did solo pilots like Saint-Exupéry or Amelia Earhart recognize this perfection of ideal form, when flying great circles over the earth's surface; computers did not fly their planes. Another metaphor, "the arrow of time," tends similarly to reduce time to a mathematical essence. Saint Augustine wrote lucidly of the tenses of past, present, and future, none of which clearly exists, but then he said that he knew what time was, until someone asked him what it was! He seems to have loosely identified time with memory, but the psychic puzzles remain.

As so often, the poets and their diurnal metaphors have an easier time of picturing time. Shakespeare, writing the finest dramatic poetry of our language, has Macbeth, perhaps his most disturbed hero, say, "Life's but a walking shadow, a poor player, who frets his hour upon the stage," where time is performed, only to be blocked by one short hour of tragic catastrophe. Thoughts of an ultimate continuum and its probable ending haunt most authors. With unparalleled eloquence Henry Vaughan, the seventeenth-century mystical poet, expressed it this way:

> I saw Eternity the other night
> Like a great ring of pure and endless light,
> All calm, as it was bright,
> And round beneath it, Time in hours, days, years,
>   Driv'n by the spheres
> Like a vast shadow moved, in which the world
> And all her train were hurled . . .

Drawing upon the astronomy of Ptolemy, "The World" is a sermon on the triviality of human illusions, from the "silly snares of pleasure" to the grand designs of "the darksome statesman," but it also captures something of the ultimate Platonic mystery, where time is the moving image of eternity. Literature and the arts attempt to give us psychological time, and much of that seems to depend on our sense of aging, which in turn feels like a natural wheel, forever turning under our feet. The rather mathematical Andrew Marvell complained to his mistress that she should cease being coy, winding life within her tresses, because the two of them did not possess "world enough and time" for reluctant dalliance or even for thinking about time, which the poem collapses into ever-pursuing rotation, where Eros vies with a running Sun, source of all the lovers' strength and all their weakness. Like other metaphysicians, Marvell imagined Life hunting Time down, only to become its prey.

Thus indeed poets, and later the philosopher Immanuel Kant, spoke of time as a fundamental "intuition," twinned with the basic intuition of space. We are not about to explain even how we feel about time passing, while the non-psychological physicist uses mathematical equations to insist on exactly the relations of velocity and time. Meanwhile, influenced somewhat by Bergson, Marcel Proust went off heroically in search of time, which he admitted he had lost. For my money—if indeed time is money—the best nontechnical book on time, and there have been several remarkable treatises, including a number (such as McTaggart's) that attempt to show that time simply does not exist, is Etienne Klein's *Chronos,* for in a curious way, as I read him, he thinks that time is some kind of joke. *Chronos* is an especially appealing study because it makes fearless connections between its scientific topic (say, the fourth dimension in relativistic thought) and the lives we actually live, with time as a psychological attribute of literature, culture, and myth, quite unlike the dimension Einstein meant, when he said that "time is what clocks measure."

All these and other epitomes and visions aside, we humans do accept a consciousness always changing, a passage vaguely resembling the technical and material and usable diurnal notion of time, when we awake and go to sleep and thereby notice there is a difference between the two states, as if waking and sleeping were two human phase changes of time-sense. Time passing is one of the most famous of philosophical riddles concerning human existence, and it remains a scandal. Indeed, in a chapter on "Boredom; or time exposed," Etienne Klein catches the puzzling *quality* of temporal passage, when he says that "boredom detoxifies our relationship to time; nothing happens except the passing of time. It puts us in contact with time reduced to the succession of moments, free of what usually contains or contaminates it." (Chronos, 2005, p. 34)

Above all, as already stated, we need to reject what Jonathan Swift mocked long ago—"the mechanical operation of the spirit." Improvement in the accuracy of atomic clocks marks an extraordinary instrumental advance, but time itself remains mysterious if we meditate upon it, and in the end there is always a "shortage of time," as of water in the desert. As the brilliant Spanish novelist, Javier Marias, has his narrator ask in *Los Enamoramientos,* What do we mean by "too late?" Perhaps we refuse to arrive "on time," but whose time is it we refuse to accept? Einsteinian synchronicity seems psychologically difficult to think with and act upon.

When Montaigne wrote in his late essay "On Repentance" that we humans sense the passing of time as if we were reckoning a drunkard's walk, he was writing at the beginning of the modern era, when peasants reckoned by units like "seven years and seven years." This was a significant historical moment for thinking about human time, for clocks were suddenly everywhere. Living beyond the clock, one explains *not* living by the clock—but what can that mean? Montaigne's essay speaks for the diurnal vision I have been describing for circadian and circannual rhythms, and it is quite unlike the twentieth-century view described by Peter Galison in his illuminating history of quarrels in more recent science, *Einstein's Clocks, Poincaré's Maps.* Neither the French nor the British could own the reckoning of time, they could not own its center, despite Greenwich Mean Time . . . because time is spherical and its nonexistent edges cannot be accorded any ultimate logical privilege or fixed starting point. Calendars and clocks reflect this sphericity, and our experience of seasons is at present markedly uncertain, given the facts of climate change.

Fortunately, by thinking of moveable conceptual edges, we can follow the observations of voyagers far out at sea, and today technology authorizes the extremely intricate analysis of climate provided by Paul Edwards's magnificent treatise, *A Vast Machine* (2010), which makes the problem of complexity clear enough; its subtitle says it all: *Computer Models, Climate Data, and the Politics of Global Warming.* Amidst the most careful and best documented overall account yet available, Edwards demonstrates the algorithmic complexity of long-term climate predictions, when he shows that we can only achieve a range of probabilities, as with quantum theory as well. "What the range tells you is that 'no change at all' is simply not in the cards, and that something closer to the high end of the range—a climate catastrophe—looks all the more likely as time goes on" (355). This is not a cheerful prospect, but the breadth of our powers of observation and analysis gives us some reason to believe that we can ameliorate our most serious vulnerabilities. We have fully entered the time of biospheric complexity, however.

Climate and geochemistry are both conditions of edge-location, but when Vladimir Vernadsky published *The Biosphere* in 1926, there existed nothing as comprehensive as *A Vast Machine,* but he took early steps to give a coherent geophysical account of this spherical process. He analyzed geology as a network of interacting biochemical bonds and exchanges. He drew upon geochemistry and the science of thermodynamics, studying the flow of heat as it fed energy (ultimately coming to us from the Sun) directly and indirectly into what Vernadsky called "the living organism of the biosphere" (1997, p. 27). Such studies would lead eventually to work like Edwards's research, where climate is an immensely complex system of flows.

The Königsberg bridge riddle had started an understanding of such turbulence. On Earth's surface, analogously, the river's delta branches are liquid edges flowing from upstream to downstream, and every island in these streams makes a nodal point or position, creating a sequence of vertices. Place constitutes a system of traceable starting points within a wider process, no matter how much bending, twisting, and stretching the watery pathways must undergo. From above, or in an aerial photograph, any delta appears as a web or meshed network of nodes and connecting bridges. Hydrodynamics sees the shaping of water.

With Leonhard Euler a door had opened onto new methods for drawing the "graph" of such complex organic combinations of networked connections and passages, regardless of size, from the organization of amino acids all the way up to various body tissues and complete animal bodies, all the way outward into the body's ecosystem. Above all the new approach allowed for a science of heat transfers—thermodynamic energy—without enlisting the animism of vitalist magic. We have compelling reasons to think more deeply about the role of those edges that make flow-patterns possible, those rivers and currents, winds and weathers, including heat-waves and electricity, and all the fixed bridges that cross them. Given such examples of crossable borderlines, we can now only go on to list at least a few prominent, not to say influential, members of the Edge Family.

VII

Notes on a Family of Edges

Attempting in his *Philosophical Investigations* to survey the variety of language-games that loosely belong together, Ludwig Wittgenstein spoke of their "family resemblances." With this metaphor in mind, let us make a severely brief conspectus of several critical edge-types, where different mental and material purposes at once appear, most of them of great importance for humans. Among properties of all edges two stand out:

an edge connects separate points (the vertices of Euler's formula) and an edge may also separate two areas (such as the inside of one country from the inside of another country). The former draws a line connecting two endpoints, as when one state boundary starts (say, on its Western side and where that boundary ends), while the latter divides areas projecting away from the line itself and linking that face with the "solid angle" or corner of each figure. (Face, surface, and side here mean virtually the same thing.) This fact indicates that the concept of shape develops eventually, after Poincaré created an algebraic approach, into a numerical method of measuring a point-source topology. What began in Euler's two great discoveries modulates into extremely sophisticated mathematics in our own time.

Lines such as longitude and latitude drawn on the surface of a sphere are entirely abstracted from the sailor's material ocean, if as in earlier times they were used for navigation. Using a sextant, by relating such lines to speed and distance, the sailor could identify his location on a blank spherical expanse. Edges in that case are geometrical boundaries of sections of the manifold. Nowadays we use satellites to broadcast such positions to oceangoing ships, and the advances still indicate edges of a larger space, reducing it to specific place. If we reduce the scale of our concern, we find the neurophysiologist examining and describing edges as minute thresholds, boundaries or junctures between different sides of membranes, as more generally with the study of cellular biology, where only by using microscopic instruments can we begin to see what actually goes on in the network of bodily interconnections (as for instance the functions of the system of blood circulation). In essence the edge articulates the system, and such concerns are topological, because the networks and channels inside the body—we might call them the biological wiring—effectively give that body its functioning shape and overall bodily character.

The list of familiar and unfamiliar edges might go on forever, it would seem, when looking under "edge" in the *Oxford English Dictionary,* but I am choosing examples that involve our planet, exemplary types for the reader to ponder beyond mere mention. For us every edge is a paradox, since Earth ideally can have no edges. The scientific and instrumental advances we have recently made have reduced the electronic world to a shrinkable sphere, but these measures of the edgeless round Earth have the further effect of electronically collapsing our ideas of time, a collapse psychologically critical, because our world-time seems to be going faster and faster. Time, not space, is the primary scene of mental endeavor. Yet edges still exist on spatial manifolds or surfaces.

Horizon

It might seem strange, and I hope it does, to begin with an edge that is not quite an edge, namely the horizon. Usually the word horizon means "as far as the eye can see," while more precisely it means the lateral line beyond which we cannot see some region of the planet, a line blocking that region beyond our visual perception. The visual array of things we see before us depends on many factors, but shows in every case that the horizon is ultimately a perceptual phenomenon. The original Greek for horizon, *kyklos,* means a separating or dividing circle, where sky meets earth, but much of the time on dry land we cannot perceive a distinct horizon-line of the planet, because buildings, trees, and irregular outcroppings of Earth obscure the dividing line itself. But what happens if we move toward that line? It moves away from us, like a friendly animal backing off when we try to pet it. Horizon, as its original Greek name suggests, is undoubtedly a most unusual scrolling line or boundary.

On the plane of movement the horizon moves always ahead of us, when we reach out to its edge, as if attempting to see beyond it. Ocean-going sailors knew this. The open sea they were accustomed to call "the offing." For them the horizon is a perceptual edge that keeps moving away from the observer, as the observer travels toward it. Owing to the curvature of Earth, one might ask whether horizon is not a prototype of all abstracted mathematical edges, which exist only as conceptual forms, floating before our minds in a free and open space? They fix only a relative edge.

It pays to consider the phenomenon of horizon at this point. Joseph Conrad's sea stories do suggest something of that experience. When Captain MacWhirr in *Typhoon* scans the horizon from the *Nanshan's* bridge, he cannot be sure which quarter of the Northeast to examine closely, so he mutters in his usual flat tone, "There's some dirty weather knocking about . . . Go and look at the glass." The falling barometer warns him of something wrong about the oncoming weather, a storm "knocking about" somewhere beyond the wide extent of a vague, remote horizontal line, for no angle of horizontal vision from the bridge will tell the Captain and the mate where the dirty weather is coming from. A blank horizon gives them no clear bearings on the typhoon, a difficulty increased because the cyclonic storm itself will be slowly rotating on its way toward them. The unknown is about to test their courage to the utmost. In another story, based partly on his own experience, Conrad describes a similar if yet more metaphysical horizon. *The Shadow Line* narrates a vicious retreating game for a young captain whose whole life shrinks to its own cyclical Vichian periodicity. He must cross a double horizon, interior as much as exterior, which Conrad calls the line between periods of unexamined youth and bitter experience, assumed success and threatening doom; throughout that tale Conrad not unexpectedly alludes to a Shakespearean drama of indecision. In his memoir, *The Mirror of the Sea,* Conrad writes of such uncertainty wrought by Nature, "in this ceaseless rush of shadows and shades, that, like the fantastic forms of clouds cast darkly upon the waters on a windy day, fly past us to fall headlong below the hard edge of an implacable horizon," and yet, on another occasion—the story being his novel *Victory,* where Axel Heyst is enamored of islands—there is a supreme quietude pervading the edge of vision: "The islands are very quiet. One sees them lying about, clothed in their dark garments of leaves, in a great hush of silver and azure, where the sea meets the sky in a ring of magic stillness." The contrasts could not be more extreme, while with another writer, whom Conrad greatly admired, Stephen Crane, the horizon may be dangerously occluded. In a way we are lost when the horizon disappears and of the four men in his story, "The Open Boat," he writes the immortal words: "None of them knew the color of the sky." Like Conrad, Crane is fascinated by the dividing horizon, in his poems as well as his fiction. *The Red Badge of Courage* dramatically shows the horizon disappearing into a general blur, for the chaos of war virtually blocks any sense of planetary edge, which leaves only the foreshortened immediate space extending just beyond the young soldier's fragile body—a body denied horizon.

Perceptually, despite its geometry, the horizon is an edge without a final or fixed occluding limit, as if its Galilean relativity never ceases, since, as it stretches sidelong at some distance along our line of sight, its position depends mainly on how high above the surface of the earth or sea we are standing. Distance perception is here a relative matter. On a high hill, like Robinson Crusoe scanning the horizon, we might see as far as fifteen or more miles, on a sand dune less than four miles, and in Conrad's sailing days at sea sailors climbed to the topmost yards, or a "crow's nest," to extend their lines of sight. Thus we may think of our being blocked by the horizon, and yet this word creates an illusion, for in fact we are simply not able to follow our earth's curvature over its edge, which forever shifts ahead of us, as we move toward it. We could better say that we are seduced by the horizon. There seems to be an elastic, even vague topological limit to our phrase, "as far as the eye can see."

The same situation pertains on land. If we drive across wide flat country (in Kansas, let's say), the boundary circle of our perception yields the same kind of rolling visual terminus, and like the fleeing runner it never quite stands still. Like the hero of Hitchcock's *North by Northwest,* perceptually as well as physically we are moving toward or away from a visual circle that retreats from our advance into a dangerous nowhere, a field of stubble, all around us, which we recognize as an ominously boundless perimeter. This condition of being unprotected is life-threatening. As already stated, the Romans therefore called the sacred bounds of a city the *mundus,* or world, to inscribe their protected and often walled civic environment. Even to speak of the *mundus* is to define a kind of dream.

Perception and belief play such a large part in orienting the science of whatever we think we know that we may forget our perspectival situation as thinkers. My favorite example is the peasant poet John Clare, who, when still a child, went off running one day in search of the horizon, before he got lost and gave up, in tears. Clare was to become a brilliant self-taught ornithologist and perhaps he thought he might like a bird outrun the distance between himself and Earth's edgeless edge. As a grown man he would speak of his territory, the familiar fens and marshes surrounding the village of Helpston in early nineteenth-century England, as "my knowledge."

Horizon and its recession are the outer (and probably also the inner) edge of all life and human perception. The world we personally know limits our horizon, and events like sunrise and sunset are prior mindsets built upon our unaided macroscopic perceptions. We are not so much wrong scientifically to think that the sun is rising, as we are imaginatively right; we can agree with the Teacher of *Ecclesiastes* or his followers, and we understand Ernest Hemingway, whose "Big Two-Hearted River" describes a flowing that forever quests its horizon. Gnomic authors like these, however different their worlds, equally raise the question of the gap between scientific knowledge located "out there" and personal knowledge located "in here"—in our conscious and finally determining relation to horizon. Sunrise and sunset ground John Clare's intense yet unanalyzed "knowledge," and no matter how abstract we may wish to be, we still do not normally picture our planet as rotating toward the sun's rays, as dawn approaches, or slowly falling away at the end of the day.

There is always a gap between immediate and analyzed experience, and the case of the horizon is a reminder that much of our belief in decisive habitual judgments is potentially illusory, and here the horizon marks a strange dividing line both *on* and *off* earth, for it seems halfway to accept the idealized fact that because spheres have no edges, the horizon promises a different world beyond, a moveable feast defining the earth's ambiguously retreating boundaries. Longitude and latitude share this function, for they mark the conceptual edges of temporal and spatial navigation over a surface-manifold, where no *final* terminus can be found. Horizon is finally an edge that says, "sometimes edges simultaneously exist and do not exist, except in mind's decision, a decision also making and marking an edge, but always within our thinking." Wisely, the mathematician and philosopher, Gilles Chatelet, was wont to speak of the horizon as a "precious hinge" allowing a sparkle to remain in our metaphysical rise beyond common, domesticated rationality.

Beckoning to our senses, Chatelet's horizon, he says, "impregnates" our sphere with the signs and far reaching signals of life.

Perceptual Edges

On the largest scale we may imagine a theoretical wholeness to our planet and then onward to outer space, and given great powers of topological imagination, it remains for the artist such as Dante, poet of *The Divine Comedy,* to convey something of the detailed texture of which our lines and points are schematic indices. Dante is famous for his precise visual images, in experience those Proustian memories that define where and who we are.

Using geometric projections our sphere may schematically be mapped onto flat surfaces, whose projected lines of demarcation include idealized edges such as the longitude. Paradoxically such edges are powerfully functional concepts or ideal mathematical objects, whereby particular locations on our planet or in astronomical space may be indicated on a map. We know that such mapping represents a real world, even when the map produces ideal marks on what might seem a striking cartoon. The map sketches the scale model of a larger whole, a principle satirized in Lewis Carroll's *Silvia and Bruno,* where the map is the same size as the countryside it represents and the farmers must lift it up, to see where they are going. In short, maps and much of topology train the explorer to notice scales and relative (though not absolute) sizes, so that that maps reach far beyond the confines of our own planet. Similarly, the great epics of literature explore an ever widening sense of cosmic and terrestrial space, at times ending in mystery, as with the metaphysics of light in authors like Dante and Milton. In more realistic literature, as in philosophy and physics, imagery is the handmaiden of perception. In this tradition Parmenides wrote his philosophic poem on knowledge and deceptive appearances, around which Plato wrote intricate dialogues, establishing the Theory of Ideas.

Maps are one thing, of course, and our actual experience is another. An image may not render what Henry James ironically called "the real thing"—indeed we may feel that images attempt to deceive us or try more seriously to depict a private perception, similar to Wittgenstein's "private language." Images leave us with a broader question, how can we know or judge if our imagery is true to some perceived fact? Can we train our perceptions? Perhaps we are born with them and need only stimulate their first infantile use.

Perceptual theory is the field where all such matters are central concerns. In trying to understand the origins of our ideas of order, we study the directly apprehensible boundaries between and within all separable objects, boundaries we find we can distinctly observe, a view that earlier psychophysicists did not fully develop. As a great nineteenth-century forerunner, Helmholtz had argued that regarding what he called the facts of perception, "in the real there must exist some or other relations, or complexes of relations, which specify at what point in space the object appears to us. Of their nature we know nothing," and he concluded that "spatially different perceptions presuppose a difference in factors." This psychophysical observation sought direct experience of the environment, but about the real perceptual "conditions" Helmholtz added the crucial reminder, "of their nature we know nothing." Such questions were not to remain unexamined in psychophysical research, however, and in particular we will shortly consider an example of experimental advance. Helmholtz was expecting to find the changing variables such as the sizes of objects, to which our perceptual apparatus would respond and neurally decode into useful information.

This new approach was to lead inevitably into a development of Gestalt thinking. It aspired to a directly realistic view of perceptual theory, a method not always reducible to mathematical expression. Gestalt approaches relate and analyze the particular image of a particular shape or texture, to a larger surrounding background. This might lead to remote questions in philosophy, such as: *Is it possible that before there is concept, there is percept?* After all, we commonly train viewers to notice certain features of the environment, perhaps by attending to the exact quality of their sensations, in which case concept precedes percept. Much of Merleau-Ponty's interest in art is meant to question the concept / percept relation, wedded as he was to analyzing the embodiment of thoughts. Our thoughts, on his view, are always to a degree psychosomatic. On the other hand, our perceptions, while often flawed as measuring instruments, belong to us the perceivers, and with that fact we are forced to begin. We are forced by nature to rely on information immediately available to us, as we move about in our world. Every particular image, every particular shape, interacts with and within a larger surrounding background—the assumption developed in Gestalt psychology.

Suppose then, for the moment, that we return to our exemplary case in visual experience, the horizon. "As far as the eye can see" might be the oldest phrase for an immediate perceptual datum, the reality of an edge, and we acquire skills in noting and using edges of all sorts, as we have said. We ask: how exactly does the visual system reach out toward the far horizon, processing the array of objects and textural arrays accumulated between that psychophysical boundary and the viewer? Children begin to acquire hand-eye skills at about the age of six months, largely because, becoming more mobile, they begin to see and feel how objects are initially defined by their edges. They move, to see. In principle, one might as well place the neurophysiology of such depth perception and depth constancy on a theoretical and experimental shelf—brain research locates different regions focused on specific mental tasks. How can all this work? As if our senses were designed correctly for the task of decoding where edges start and end, their adequacy in processing the visual is not something that can be replicated easily under controlled laboratory conditions. The "conditions" Helmholtz referred to appear in all areas of a surrounding world, and their belonging participates in the ecology of perception.

To pursue such a vision was the destiny of one remarkable perceptionist. Over many years James J. Gibson at Cornell University pioneered study of what he early called *The Perception of the Visual World,* the title of his first book. Significantly, Gibson was fascinated by works of visual art such as paintings and film, and especially with the former he was aware of their inaccuracies and their accidental flourishes. With his colleagues, among them the distinguished expert in child development studies, his wife Eleanor Gibson, he focused attention on perceptual powers exercised in *the real world surrounding us,* our living ecological surround. Most of his final thoughts from this perspective are presented in his late text, *The Ecological Approach to Visual Perception.* A legendary story goes that he became interested in the "visual cliff" phenomenon, whereby virtual neonates would instinctively back away from the appearance of a dangerous declivity (reproduced in the lab at Morrill Hall), because one of his own children had crawled away from the edge of the Grand Canyon, having looked over the precipitous drop. Something at the South Rim was triggering an aversive response. He asked how perceiving the visual array of any perceived texture was objectively related to movement, movement being the chief non-laboratory condition for living creatures, and he noted motions in the scene presented to the eye through movement in the surrounding world observed under common conditions. Motion here could mean the movement of the observer's eyes (e.g., the vision of a walker or a bus driver or a fighter pilot landing on an aircraft carrier deck—the specific mission of Gibson's earliest wartime research) or it could mean the contraposing shifts of position, large or small, in "the world out there."

Motion and perception interact, and this principle animates the developing Gibsonian theory that perception implies an ecology active in our experience, whose principles of direct (if perhaps hazy or blocked and occluded) perception are explored in the 1979 book outlining his ecological approach in great detail, following a number of related technical articles. The idea that ecology as an evolutionary component implies an adaptive evolutionary role for perception would appear almost irresistible, but that would be a much larger complementary question, Do our senses allow us to be continuous with nature? Do we learn by repeated contacts? Gibson answered his own question topologically: "The information to specify the continued existence of something may be carried by touch or sound as well as by light. Incessant stimulation is not necessary to the perceiving of *persistence*" (*Ecological Approach,* 208; Gibson's italics). As already noted, the complex picture of such events depends on various types of eye movement, for example in the way the head moves and also in minute saccadic eye-motions. Binocular vision itself implies a motion adjusting the angular difference between what the two eyes see, although, by moving the head, a person with only one eye can reproduce the binocular effect. In this context *constancy* means topological invariance and "the perceiving of persistence."

To perceive an object's boundaries also means that we, like other animals, can tell where we are placed, despite an ever-shifting world, and we use this permanence in change to orient our actions, while the entire Gibsonian approach emphasizes control of individual locomotion and manipulation of *placing-data* (without any Heideggerian overtones, he calls it "way-finding") that contributes *directly* to the perceptual powers of the seeing person, when the perceiver actively processes an array of information.

It would be unreasonable to compress Gibson's ideas and the massive detail backing his theories, but a topological account will highlight one central issue. For acquiring a sense of the environment, a variety of objects need to be perceptible: *place, attached objects, detached objects, persisting substances,* and *events.* Particular things perceived are not, however, entirely individuated. The *Ecological Approach* suggests that instead of telescopically *seeing* our environs, we *view* them, as if we had not yet discovered the geometry of Renaissance perspective, which in turn implies that rather than calculate distances and sizes, our perceptions directly gain visual information, but without calculation of point-like visual signals, as would be necessary for any *signal detection* approach. In the process of this coherent *viewing* the externals almost become a part and parcel of the perceiving self—the author actually speaks of a *co-perceiving of the self.* Ecology, Gibson notes, justifies our focus on ambience, on ambient light, on our being surrounded by "arrays" of stimuli—all of which enter our perceptual systems as externally organizing topological displays. We recall William James's remark that discrete objects ejected from the stream of consciousness are "as mythical as the Jack of Spades." The irony is, however, that we humans and other living creatures are never adaptively free from the need to register edges.

For ecological purposes, as similarly they involve imagining and not just observing where we are, the most telling point of departure may be a short passage Gibson devotes to *place.* As already noted, in the Gibsonian approach we depend entirely on processing the virtually universal fact of the motions of things, events, and scenes, all of which become useful or apparent to us because we exist at various levels of motion. Place-movement or locomotion is critical to life. Commenting on our two-dimensional manifold—the Earth's surface, which is virtually the most general term we have for ecological space, and recapping his Chapters 1–3, Gibson says:

> A *place* is one of the many adjacent places that make up the habitat and, beyond that, the whole environment. But smaller places are nested within larger places. They do not have boundaries, unless artificial boundaries are imposed by surveyors (my piece of land, my town, my country, my state). A place at one level is what you can see here or hereabouts, and locomotion consists of going from place to place in this sense (Chapter 11). A very important kind of learning for animals and children is place-learning—learning the *affordances* of places and learning to distinguish among them—a way-finding, which culminates in the state of being oriented to the whole habitat and knowing where one is in the environment. (1979, p. 240)

Thus, to return to my earlier pages, place *(situs)* and placement in the original Leibnizian sense are central notions for Gibson's approach, but I have italicized *affordances*—a key innovative term for Gibson—because they comprise all the perceptual features of the habitat which *afford* perceptual entry into the surrounding environment or place. In a Helmholtzian sense the variables of shape, texture, movement, orientation, and so on all combine to *afford* critical conditions of perceiving correctly. The environment is not always favorable; it is visually often harsh, as in deserts, and in such dangerous parts of the world, where perceptual cues may dissolve the whole array into a featureless *ganzfeld,* it does not afford any opening to Gibson's "way-finding," nor may such ecosystems provide sustenance. Nonetheless, as humans navigate their world, for survival alone (in every sense) we find that an ecological approach to perception is critical at every layer of the biosphere, all the way from subsurface water and microorganisms to the upper reaches of the stratosphere, since we are now able to examine all these layers—ironically, at the moment in history when we humans are interfering with their complex integrity.

In terms of information also, we find that while we wrench reality from our sensed surroundings, those surroundings remain our chief informing resource; ecologically they *afford* us masses of data we directly pick up and then use, to perceive the environment. In this aspect of the ecological approach we are reminded, as elsewhere, that Gibson was much influenced by the third great Gestalt psychologist, Kurt Koffka, who described an environmental "invitation character" which displayed a perceptual invariance relied upon by humans. As Gibson calls them, the environment displays particular "niches" opening, refining upon an otherwise cluttered array of stimuli, restricting an excess which would otherwise overload the perceptual system. The niches organize what would drift indiscriminately into a picturesque heap, an impossible *sorites.* The *affordances* permit the seemingly limitless detail to be curtailed or focused, which in turn allows perception to move from unbounded texture to created shape, liberating the topology of the scene. This topological approach enables us to find the invariance in the unordered wider array, as when children discover edges of things they might bump into, and their explorations cease to be random lunges hither and thither.

A concluding Appendix to the *Ecological Approach* specifically lists and defines four types of invariant, starting from the principle that "the theory of the concurrent awareness of persistence and change requires the assumption of invariants that underlie change of the optic array" (ibid., 310–311). For example, there will be invariants of optical structure under changing illumination, but there will also be invariants of optical structure when the target object is seen from a different point of observation. Of interest to the artist would be the invariance occurring when the object is "locally" disturbed, as with a breeze blowing through the leaves of a tree or flower. Virtually all living creatures pick up on any local disturbance, as when registering wave motions in a tumbling stream. We all notice changes of form, instinctively.

Yet our powers of discerning change are not easy to explain as perceptual achievements, even though physics is deeply engaged in analyzing many forms of motion. The distinguished Chinese systems theorist, Luon-an Chen, wrote in an early paper, "Topological Structure in Visual Perception" (1982), that "we now do not know how this resonance to topological invariants is generated. This kind of question lies at the core of issues in perception" (699–700). Basing his thought on the topological work of E. E. Zeeman and using tachistoscopic displays, Chen argues that Gibson was right to say: "the perceptual system simply extracts the invariants from the flowing array; it resonates to the invariant structure or is attuned to it" (ibid.). Were he alive today, I suspect that Gibson himself would hold that we still know much too little about the way the mind and body achieve the topological resonance with invariants, as if Euler's Polyhedron Theorem yet remains a miracle of theoretical constancy that appears "all over the place," to use common parlance.

We can only make informed guesses about the complexity Gibson discerned, as latterly he analyzed what happens in perception. For example, we discover that somehow any "occluding edge" will not totally block us from estimating what lies behind it, when we move toward or away from it. (Looking out my low front window, which faces a paved road and beyond it a park, I count more than twenty-five distinct edges before I reach the clouds in the Eastern sky, and I wonder what I must be missing in the count.) We discover that every edge is the sum of an infinitely more and more detailed, smaller and smaller set of sub-edges, as the very different world of air-traffic control shows that planes follow distinct paths in space, but these need to be clumped or *bundled* together for matrix analysis, yet without totally destroying their detailed differences, from plane to plane, in direction and speed. Edge, in short, is an abruptness of change in texture, but texture itself seems edgeless when arrayed before us. The artifactual as well as natural world thrust upon us the idea that edges and surfaces are infinitely multiple sheaves of almost pure difference, which we still need to study in all their multiplicity.

No contrast could be more stern than the difference from this multiplicity drawn by Euler's Polyhedron Theorem, where we deal with three entirely abstract components, the Vertices, Edges, and Surfaces of the Platonic solids and their myriad derivative shapes. The invariance summing those three features is a remarkable feature, so purified that it appears almost unnatural. Like other scientists, however, when looking at the perceptual constancies (size, shape, color, and so on) and their method, Gibson was in fact thinking of the Eulerian invariance, but he knew that real-world visual perception imposed radical changes lying outside the purity of the Theorem. For example, the illumination of the visual array could change the task of identifying the reality of the perceived edges in a landscape. Hence in his *Ecological Approach* Gibson found difficulties with reliance on a purified geometric approach, or any purified topological invariance.

His hesitation regarding the life-world is exemplary, and poses a philosophical question about abstract thought modeling biotic arrangements. The legacy of his career thus in general calls for a radically realistic approach to the workings of mind, since, to invoke the spirit of Merleau-Ponty and Professor Chen, the mind is an embodied attribute of the larger environment, whose ecology is known or experienced by us insofar as we humans, like other living creatures, inhabit the world *through* our own bodies. Inside and outside finally must possess each other. In the final analysis objects cannot be ultimately distanced or perfectly divorced, so as to lie "outside" the observing mind, no matter what materialist myth is invoked in today's world. The term "perceptual" of course refers to our interface with the world, and that interface is an inter-action, moving across a bridge or threshold to the possible description of what is real in a material frame of reference. There is no reason to be complacent about knowing exactly where and how the cut is to be made, and then what we should make of the liminal division, once we have achieved our cutting edge, since every edge leads to another finer one, as we know from sciences like spectroscopy. If there is a special emphasis for method, from Gibson's example it must be that we live in motion and motions of infinite variety, as if the edges kept on growing and decaying, and so we are left to contemplate invariance amidst change—perhaps the chief property of topological thought. The impression we are left wondering about is the obscure meaning, once more, of our division-ridden world, where boundaries are an almost universal obsession.

Cutting Edges

To cut is to cross a line, when we speak of "cutting over to that road instead of this one we are on," while "cutting to the quick" means . . . to be incisive when penetrating to the heart of the matter. The problem is to draw a clear picture of what happens in the space between the material and the mental. In a psychological sense we understand incisiveness to be inventing or using *concepts* that precisely mark borders between this thought (on whatever subject) and that thought (on the same subject). The ultimate cutting edge is always a clear distinction drawn between two or more ambiguous possible meanings, and hence is a matter of thinking, of logic, of cognition, and only then of a decision regarding some proposed action. Even when the issue at hand involves real-world consequences of grave importance, the mind is asked to do the *cutting.* Many good people refuse to make distinctions for fear they are being prejudiced. How can we then *think* about our global need to understand the Earth's ecological condition, since without studying the action of edge-making and edge-decoding, we will never understand why serious ecologists often say that the Earth is at a crossroads and has no more time for delay in reducing, among other hazards, the emission of fossil fuel effluents such as carbon dioxide into the atmosphere, the polluting and plundering of the oceans or the organic resources of native soil. And yet the immense scale of these changes virtually defeats our thought, for we continue to exploit natural resources and meanwhile allow world population to increase exponentially. Scale alone baffles reflection, or encourages wrong-headed short-term solutions. It is possible that the shape of the globe itself is defeating us.

The issue for acute separations of one thing from another remains a conceptual one, and a cutting edge has always this conceptual, nominalist role to play. To study the cutting edge is not, I hope I have indicated, a mere postmodern game of "decision theory," which seems always to return to the dollar sign as its main term. Though world populations may have already gone over a limit and cannot undo this loss of control, we may even then be getting the wrong idea if we imagine a catastrophic crossroads, the wrong idea if we think we have gone through a red light and have earned a life sentence for the violation. This analogy is not appealing, because, while the jury is restive, there is no real world court, there is no judge to pass sentence, although it does suggest that while the earth's climatic emergency is every day more obvious, we still need some care in imagining our global plight and policies. We need to revise our attitudes toward what we consider "realistic" over the long term, for we know that to progress we must cut down on excess.

Already we have seen in our previous comments on Gibsonian perceptual thought, that ecology provides the most general and most commanding frame for sensing global imperatives. In essence what is real is what entirely surrounds us, an ambient sphere, in fact. By supposing that we can take the notion of edge still further, there could exist a method of dealing with what is so utterly disturbing about our environmental crisis, namely the apparently absurd scale and complexity of the organization involved. Nothing less than a conceptual revolution will suffice, if we are to find the necessary scope of our inquiry. That the planet functions for life as a result of self-organizing feedback loops is not merely plausible; it is the only useful framework for explaining the evolution of life-forms over vast stretches of time and space, and certainly its violation fits exactly the unwanted side effects of much of our human-centered technology.

Hidden in the idea of edge there is also the connected cosmic motion of all matter, which calls for a new attitude, moving our thoughts into unfamiliar territory. The story of motion is the story of all physics, which concerns the perception and measurement of the movements of the smallest to the largest material objects in the universe. For every edge may be conceived as an artificial dividing line between two stable states in time and space, drawn as if the line itself were not always moving, a static border remaining a boundary though still capable of disturbance. In a world largely determined by fields of force, such as electrodynamic or magnetic effects, we may forget that in a radical sense, that is, topologically, every edge can mark either the beginning or the ending of some area or patch, and what interests the ecologist locally is the way we respond to conditions occurring within such an edge-defined patch, some patch of some surface that, through our observations, *begins* at one or more of its edges.

The issue is one of identity carefully conceived. Whenever we cross a border to the other side of the separating edge, we gain the potential of thinking anew about what is static and familiar, usually about our own familiar homeland. Borders are always to some degree uncanny, as travelers know, for we are strangers "coming from somewhere else." We have to realize that data accumulated in science are also always coming from somewhere else, a somewhere needing to be related to the middle scale of our observational powers. By using the idea of the conceptual cutting edge, we can reduce the sheer numbers of large populations to statistical analysis, and hence in a sense we can translate the numerous mass to a sort of set, a complex arrangement of agents or actions. We are enabled to ask who "we" are, if we can criticize ourselves as members of a group. This combinatorial arrangement (our own version of the Bridges of Königbserg) produces an organization of shaped totality, in which we humans recognize our shared, enfolding, environing world. We (but who is this "we" I am forced to speak of?) always belong to a controlling larger manifold. In this view, as we humans contemplate where and therefore how we live, we are neither exactly alone nor exactly conjoint social creatures; we are infolded beings, parts of a larger metonymic system. The edge that on a topological level is absent from a true sphere appears in nature, finally, as a line where folded edges promise harmony and growth. Our manifest destiny, as it were, needs to become respect for these extensions of planetary surface, including those of the oceans and the air we breathe. Abstractly our need is to cross the edges in nature and artifice, but in that need we find the contradictions of our quest, for at once, thinking this way, we discover what Doctor Johnson in his *Rasselas* called "the insatiable hunger of the imagination." In the world we are now busy constructing, Candide is no longer so free to tend his garden, having seen the world in all its folly. The garden was the original temple, but now it is surrounded by a labyrinthine wilderness of insatiable hunger.

The simplest way to imagine the psychic edge that demands escape or at least a saving movement is to call it a threshold, something like a doorway between the inside and the outside, seeing the lintel as a frame for a moment of change. Such a dilation, what we come to realize, is a paradox of a ritually slowed passage into knowledge and experience, not entirely unlike the deliberately slowed "rites of passage" that mark all successful sequences of learning. The play of mind in such learning is largely a game of determining what Jeffrey Weeks's 1985 book, *The Shape of Space,* discovers in the topological discipline of continuously shifting boundaries.

I speak of *cutting* edges, because the central notion behind this whole list is that of a mental and also emotional response to the world, a response that cuts into our unexamined lives, to find our true boundaries. If that desideratum seems abstract, so be it. But we do need the cutting, the incisions in our thoughts. Having imagined many cosmic designs and models, we are free to build their vision into real things, full of wonder and delight. Through an exactitude of shape, this mixture of the real and the ideal happens in science and art, all the time—just consider the import of the Swiss Patent Office for the 1905 *Theory of Special Relativity.*

Constructed Edges

The most obvious paradox regarding edge appears when humans build actual structures on a surface which in principle has no edges. Everywhere in the natural world we find boundaries, but the mark of civilization is to construct boundaries for material reasons and for what we call real purposes.

The rivers may flood their banks in London and Paris, Prague and Petersburg, like so many cities the world over, but their embankments and dockside walls resist the flooding; harbors and beaches cry out for breakwaters; dams and sluices everywhere oppose uncontrolled flooding, so that cities may come together to make an urban center—everywhere we find that we have built barriers depending on sturdy anti-natural edges. Inland there is scarcely an imaginable dwelling place built without corners or rounded edges, no matter how varied are the materials used or the aesthetic of a particular design. Curved edges make a striking effect on the eye when suspension bridges lift the center of the long span into a gentle arc, and at least one reinforced concrete bridge designed by Maillart actually turns its roadway to meet obliquely angled natural edges of a gorge in the mountains, or the cantilevered rectilinear edge of Wright's *Falling Water* house as it deliberately contrasts with the natural curvature of a liquid cascade falling below it. These may be considered exemplary aesthetic cases among thousands familiar to the historian of architecture, whose classic field of study perforce lays heavy emphasis upon a small number of surviving Greek and Roman temples, where the monumental designs, based on rectilinear and sometimes circular geometric forms, bring out the contrasting fuzziness and twisting irregularity of natural forms. The Corinthian column is an exception, for it treats the capitals of columns in a shape derived from branches, although even here the column gets its load-bearing tubular shape from a topological transformation of edgeless torus.

Domestic living spaces, homes of every kind, from primitive huts to perfect tents and igloos to elaborate many-roomed palaces in the baroque and mannerist style—they all tell a story of finding the perfect edge for the builder's purpose, by adapting to what was available or through discovering new building materials. In the old days it was common to meet carpenters building houses by hand, let us call them Bob and Hawley, who built a house in Connecticut from a scrap of paper with a pencil sketch on it. Estimating by eye, they could trim the edges of a beam, knowing where one edge would meet another on the sloping roof. But we need not gorge on handicraft examples, for the constructed edge is essential to all machinery, from James Watt's famous tea kettle to the Large Hadron Collider at the CERN facilities in Europe. Industrial production depends entirely on machined edges, as the plastic lens inserted in the eye, after cataract surgery, needs to have its perimeter exactly shaped if it is to unfold inside the eye, to assume its correct orientation and focusing power. One can hardly imagine our world without such precision in use, so there is no limit to this cataloguing task. Wherever we contemplate our craftsmanship and technical advances, the precise measurements must be precisely engineered in production, which depends finally on our recognizing the art of the edge, where technical control ensues from the correct drafting and critical concept. This means we attend to more than the mechanical drawing involved—we attend to its practicability.

For there is no doubt that edge is a conceptual mixture of all the senses cooperating as some final production of the idea, which in topology is understood to be the line connecting two vertices of a regular solid, such as a tetrahedron, which gets its name from the fact that it has four faces, or the more complicated icosahedron, which has twenty sides. In this context edges do more than connect the sharp points (or as Euler called them, "solid angles") of the shape; they create what we call the "sides" or "surfaces" or simply faces, as when four straight lines enclose what we would call "the side of the house" or "the area of the cornfield."

Edges, in short, can model material enclosures, and when early modern Europe saw the end of the traditional common lands, hedgerows and lines of trees were used to mark edges of the newly bounded private lands. These bitterly contested hedged properties were called "enclosures," a term applicable to vast territories throughout the world, some still tribal in administration (if not real control), many sequestered for military use alone, others for commercial capitalist development in agriculture, and others for public use, such as national parks or conservation programs. All of these enclosures imply self-evident uses of land and sea, where various rights of possession are asserted.

These rights and prohibitions are all human constructions, as it were the schematic blueprint of political and social reality, and the term "construction" here takes on an extended political meaning, when people decide by human social covenant what will or not be enclosed by the edges called "borders." At such historically produced junctures the edge delimits sovereign power, private or corporate ownership, or other similar legal divisions, separating an extended territory into a shaped private domain, a face in the topological sense. When a political power asserts the right of eminent domain in order to promote some special public interest over previously accepted private ownership, the law is being used to construct real edges, and we know that laws are forever rewriting edges all the time. That is what happens when political parties manage to gerrymander voting freedoms, which can shut down governments, through deliberate plutocratic disregard of democratic process. In such cases, many of which have caused the most violent large-scale wars, the central constructive aspect remains conceptual (often called "ideological"), which is hardly an academic exercise, considering the lives and blood shed in the interests of the claims involved. If Nature does not insist on providing edges to control life, power-hungry beings will certainly attempt the takeover. They will construct it. Nor is the construction of social edges in principle devoid of reason—on the contrary; there are always reasons invoked to justify redrawing the lines. Our sphere is a special shape of object in which our constructed boundaries are always forced to deal with a most irregular spherical surface. Here, recalling earlier "enclosures," I mention political structures and the abuse of ownership, solely to emphasize that permanence in change is a legitimate social concern for the ethical implications of topology.

As always in relation to human ingenuity, construction here will mean many things in relation to the sphere, not least where topology builds laborious distinctions between boundaries and the unbounded. Here the imagination calls for an idea of the manifold. Abstract thinking must serve as a method of classifying things, and as Professor O'Shea observes on several occasions, a universe (including the unity we find in our own planet) can be bounded or edged, while that implies nothing immediately in respect to its material constitution, nor its measured extent. Furthermore, this a matter of correctly modeling the shape of our planet, treating it mainly as an ideal mathematical object, which can more realistically treated in, say, a suryeyor's perspective, as an ideal shape which is actually deformed in topographic details, such as hills and valleys (2007). We make physical and ideal models in order to conceptualize. The idea of our sphere as a two-dimensional manifold may seem odd, but it is intended to guide thinking in a certain direction, toward understanding the implications of its shape, rather than cataloguing its content, and for that reason the topological sphere is solely the mathematical notion of the outer surface, leaving quite unlisted all the innards, the solid three-dimensional "ball" of our planet. Our sphere is thus imagined solely as a surface, until by metaphor in its most violent form we may break through the ideal, Platonic perfection of the sphere as geometric construction. We now understandably imagine the Earth as a gigantic round ball, which in one sense it is, but on the other hand the pure form of the sphere exists on a higher level of abstract thought, the mathematician's idea. However, the force of gravity originally was an eccentric artist. Gravity forced the round ball to take on an irregular material shape, flattened at top and bottom, roughed on the surface with dips and hills, canyons and mountains, whose material departure from the ideal cannot cancel the truth of the ideal geometric object.

What are reclining chairs all about? Maybe, as its designers intended, and I have learned firsthand over the years, the physical theory of the recliner is what carries meaning, while there is meaning in a motive, surely. Consider the beaver dam or the ant heap or the bird's nest or the dung beetle rolling its cooling food supply along a blazing hot path in the South African veldt. These are self-protective devices.

Yet we humans have insatiable minds,not to mention our instincts and desires. Consider human dwellings and forts and terraces and a seemingly infinite variety of local boundaries designed to foster survival. Eighty-story skyscrapers may be conceived and engineered as machines for sheltering bureaucracy: anthropologically elaborate private residences turn families into bureaucratic communes housed in the tallest buildings of Dubai. Today the *sides* of extremely tall buildings tend to provide their main support against wind and weather, as for example the new Shanghai Financial Center building—the world's tallest, for the moment at least. At times one cannot help wondering if the desire to live and work inside a vertical shoebox is not the strongest of all human passions—until the elevator malfunctions, or worse. What seems to happen with the use of constructive edges is that our technical skill, extraordinary in itself, manages to defy earth's implications—here, defying gravity, by exploiting its downward, earthward force—as if to prove our human limitations by artificially exceeding them, and this defiant attitude is clearly emotional.

We just want to construct things that have more of some property than the other man's building, and we are good at using edges like the edges of cardboard in a shoebox, to carry stresses. With all this activity we sometimes lose sight of nature, and take too little instruction from her wayward forms. On the other hand humans do start wars with nature, when digging ditches for irrigation or levees to control rivers. What after all is wrong about building these new edges? Probably nothing, so long as they do not undo their own reason for being. The trouble is, the higher the breakwater, the greater the advancing wave. The faster the automatic weapon, the tougher the bulletproof vest. Action and reaction dominate the art of war. Poisons breed antidotes; medications have innumerable side effects. There always remains a measure of progressive uncertainty, which science may observe yet must remain uncertain about, because science advances by constructing yet one more clever theory, to account for one more cleverly constructed uncertainty. When it seeks to enclose, the constructed edge finds that it must also divide.

Natural Edges

We have said that edges and other topological forms are conceptual in mathematics, material in our ability to use them for building, and we now turn to their observable natural existence. A useful bit of word play may help us here, since the geometrician "constructs" his abstract Euclidean figures, while the builder "actually constructs" what he reads off the blueprint; he materially crafts a thing corresponding to the design. It is not easy to distinguish constructed and natural edges, for conceptual reasons.

Architecture, however, closes the gap between made object and conceived design, and we intuit what Goethe meant when he said that architecture is "frozen music" or what the ancient Greeks (and Sir Philip Sidney in Queen Elizabeth's golden reign) meant by the architectonic shaping of poetry. Which action of mind freezes the motion, we ask? Construction, real or imagined, seems equally an imposition upon nature, quite opposite to the fractals and "roughness" defined, however, as distinct formations in the Mandelbrot Set. Close observers of things happening outdoors often report sensing that Nature is a wild, inspired architect, and to take a typical case, we find constant shifting whenever we study shores and beaches or wide natural growths of marshlands. This vagueness of edge makes such natural edges vulnerable to human predation, for instance in the interests of developers. The edge itself may be only an approximately determining line or cut; it cannot be geometrically exact.

While engineering and conceptual design are close in spirit, we must take a bigger jump as we encounter the rich world of living beings, on every scale, since in this scene of natural topology the shapes of life just happen. Leonardo da Vinci's *Notebooks* provide startling line drawings of such natural edges, as when he depicts a deluge or a wild storm. That we can explain how things just happen is the aim of science, of course, but the happening is not in some primal sense occurring because we humans designed our sphere with all its complexity. That is what we need to discover and explain. That we need to imagine, and there is no point in trying to get rid of the idea of nature, though it seems an idea inviting some to engineer it out of existence—unless, with some futurists, we do believe things would be better if we and the rest of it all were just machines. Tidier, yes, but "better" raises a metaphysical question.

What I should like now to say is that the natural edge is the basis of a coherent biodiversity "decorating" the coherent order of emergent life, when decoration and structure merge, two aspects of the same process. For life and the biosphere are less consistent in their multifarious interactive phenomena than they are coherent; they comprise, as a whole, a manifold of cohering connections, whose networks are never, indeed must never be, orthogonal like a checkerboard. Contemplating the way our biosphere holds together, unless we destroy that holding, we find that we need various different visions of what such unifications actually are or do. Coherence and consistency are not similar in spirit, for the latter is only a formal game of following certain rules. Strict consistency, if followed according to classical If-and-only-if logic, blocks life from emerging through chaotic phase-changes, where, as with ice floes turning into water, turbulence generates different orders of coherent organization—as the incessant falling of waves and slicing of tidal currents brings changing species of different creatures into a reproductive proximity. On the other hand the powers of statistical mathematics have been arrayed, with remarkable power, to describe what traditional logic cannot precisely follow. My intuition is that such descriptions leave unmet the critical task for humans, who must actively deal in crude approximations, and nowhere is this difficulty more apparent than with the formation of natural edges.

All by itself, we may say, the natural boundary makes unexpected changes possible, or rather, at the natural edge we discover the complexity of natural forces operating on free bodies, such as clouds in the sky or stones, shells, sand, or seaweed at the edge of the sea. Nature herself shapes each rough collision into an ever-changing border zone, and ecologists have long sought to converse with inhabitants of these boundary conditions.

With natural edges it is as if the horizon walked toward us, to greet us with the news of perpetual change. Nor in the event must such encounters of the far and near be translated into a purifying geometry, but rather they tell us that such an approaching distance is all we humans ever have, when we engage to meet the actual world around us. This approximation of the boundary, this sudden awareness of the strangeness of distance, this danger that comes when we actually notice things up close and see them for what we think they are, all must occupy the attention of anyone trying to imagine nature's edges and their connection to life.

These boundaries may be treated mathematically, but today in a time of catastrophic climate change, it pays to consider the experience of actual observers and inhabitants. Earlier, of course, we could read the best natural history writers, for example Thoreau's *Journals, Walden, Cape Cod,* and other texts. For my purposes, however, no writing has more clearly rendered life at the edge, with its always shifting local biosphere and its "disparity actions," than Rachel Carson's wondrously detailed first book about her journey into the marginal, which she entitled *The Edge of the Sea.*

While best known for her attack on the indiscriminate use of insecticides, Carson was trained as a marine biologist. In 1948 she was engaged by Houghton Mifflin to write a standard field guide to the Atlantic seaboard, but found its genre was not working for her and she scrapped the publisher's original plan. What followed from her decision is well worth emphasizing in some detail.

Carson used her experience as an explorer and skeptical geophysical interpreter to reflect on the liminality of what her first chapter called "The Marginal World," where land and sea dilate the threshold of what may well be our most important ecological edge. Personal experience plays a key role in such research, I believe. I myself remember playing around the edges of tide pools in the West of Ireland, long ago, fascinated by sea anemones and tiny crabs the size of a thumbnail, and later for many years lived on Gardiner's Bay across from the Island given by King Charles II to Lion Gardiner, as I recollect the history, so that I am not quite an objective critic of what Carson was doing in her research up and down our Atlantic coastline. To a child tidal pools and the wetlands of Accabonac Creek were a fractal world, long before anyone heard of the Mandelbrot Set, and what we gazed on before the tide returned was a drama of danger, suspicion, attack, and escape, where tidal currents shifted invisibly across an infinite transparent stage. While my own early experience roughly resembles that of Rachel Carson, it completely lacked her scientific discipline as a naturalist.

Her observations in *The Edge of the Sea* remain so moving precisely because she makes no secret of the underlying predatory drives of the search for food among creatures of the sea and shore, and this willingness to confront nature's wild barbarity marks her research as basically metaphoric, as so often time is momentarily suspended, like the jazzman's stop time chorus, like a moment in a much larger music. In a letter to her editor, Paul Brooks, Carson wrote that she was planning "a biological sketch." Almost as if punning on the word "biographical," she wrote Brooks that her sketch would in every part suggest "a living creature," an uncanny anticipation of the terrestrial expansion pioneered by James Lovelock in his Gaia hypothesis. Carson at first seems like a modern Romantic, but for every single creature she wished to illuminate how it survives in its environment, by showing where its food is found, where it must fight with comopetitors for that food, how it finds a community of associates. How different and less genteel is her post-Darwinian tone from that of Gilbert White's 1789 *Natural History of Selborne,* and yet she follows clearly in White's early ecological trail. They both belong to the tradition of Natural History. There is nothing genteel about the meeting of land and sea; life and death are partners. The beauty of the chain of being depends upon the energy of a savage instinct to survive, and once again we must learn to accept colliding opposites, as language demands the metaphoric discords we have already found in Nicholas of Cusa, as in Shakespeare or Andrew Marvell, or for that matter in the writings of Bartram, Muir, Lopez, the late Peter Matthiessen, and many other "nature writers," including environmentalists and activists such as Bill McKibben, many of whom have been expert and exact observers of changing natural conditions occurring in the wild.

With her last book, *Silent Spring,* Rachel Carson broke into the front ranks of science writers and she changed public opinion forever, if not sufficiently to change the practices of companies like Phelps Dodge, Dow Chemical or Monsanto, or the use of *Agent Orange* in the Vietnam War. Her attack on unrestricted pesticides may benefit growers and city dwellers in the short run, and also for plain comfort, but in the long run may irreparably damage whole ecosystems. She became a hero of ecological consciousness, and while admirers and the usual opponents have found the attack too simple, *Silent Spring* remains a major manifesto, although the interests of war seem always to trump any other aims of what is uncritically called "policy." Sue Hubbell, commenting on Carson's style, remarks that the finesse of the earlier book owes much to its more acute personal meeting of author and zoological life-world; the edge of the sea is home to old embittered barnacles as much as to tiny mysterious ghost crabs; the margin between land and sea pullulates with struggle, but also with successful ant-colony coexistence as conceived in E. O. Wilson's expert yet idealizing biodiversity.

For the edge of the sea is a model of what ecologists call an *ecotone.* It is the scene of edge effects, where despite disastrous human interference with natural rhythms, serendipity abounds and life increases in power and variety precisely if shoreline developments leave these edge effects alone for considerable periods of time. For example, researchers at Brown University have found that the Cape Cod ditching programs of the 1930s, designed to get rid of mosquitoes, along with certain fishing practices, have left damaging results almost 100 years later, and furthermore these historic intrusions have synergistically combined with real estate development to multiply the long-term damage. On Long Island I remember the "Mosquito Commission" only too well, because where we lived, looking across to the old white wooden manor house, before it burned to the ground, my own mother (she grew up swatting midges in the Scottish Highlands) was forever praising men pouring oil on pools and looking for rotted tree trunks where mosquitoes might breed. She was extremely sensitive to mosquito bites, although a tough person in many other ways. Our relentless human pursuit of the better life was part of a county ditching program that was later found to be destructive to the environment!

The critical point about noticing still unchanged natural edges such as the beaches of Cape Cod or Long Island Sound or the great coral reefs on the other side of the globe is that when we interfere with edges in nature, they can no longer contribute to nurturing the true complexities of ecological action. You can take the action, admittedly, like diving ten feet down to retrieve a waterproof flashlight that slipped overboard, because you can see the light shining dimly next to a barnacled piling down below—but you hesitate when you think of the large crabs just waiting for your fingers, for their crab salad. Best dive quickly, and then leave things alone. We go on the attack with undue relentless force, it seems, as if we could eradicate our competitors without any cost. Life at the edge of the sea, like life on other edges (for example, gambling in any form), calls for sensible far-seeing discriminations. That is the task set by the ecotone, as we think of the edges separating areas marked by clean-cut ecological difference, for example a shift from field to forest.

Discrimination is the conceptual issue because nature does not always present such a dramatic or threatening face as the edge of our North American coastline, or the Norwegian fiords, or the sudden South African splendor of Table Mountain. Edges of geophysical reality are more often infinitely delicate, as science leaves the wonders of natural history and turns to statistically significant measurements in the subatomic world. We now measure reality in wavelengths of radiation, using spectroscopy, whose instrumentation permits registering and hence seeing differences of wavelength unimaginably small, unimaginable, that is, unless mathematically we are comfortable with Planck's constant, h, which is so small that no one ever noticed it before Max Planck's discovery. (But then it may be that physical constants are the last to be asked to dance.) Our recent computerized instruments are so powerful that they bring us information completely beyond the ordinary mid-scale world of human perception, as I have already noted, but this is no way diminishes our need to discriminate which are, and which are not, the significant data. A serious theoretical scientist learns to make such judgments, partly through the individual's understanding that judgment is never a dispensable skill. One practical reason for this requirement is that our new equipment is unbelievably expensive to build and to operate, not to mention one needs an army of trained experts to use the devices in an efficient way—or so the pragmatic bottom-line argument is always presented to the lay public. Playing on the old Latin saw, who is to judge the judges in our newest sciences?—for one could argue that without imagination refined through a sense of metaphoric disparity, there will be no solution to this paradoxical question.

On the macroscopic level of ecological research such disparities are centered on the changing scene of the ecotones. A naturalist like Carson may seem too individually committed to her own personal quest, but without such personal engagement there will be a serious lack of insight into the dynamic of natural process, with its fragility as well as its power. The trick with discriminating correctly is to let the mathematics and the method have their own passion, yet still maintain their judgment—as the greater Romantics always said. For this balance to occur there must always be a method of conceiving the appropriate boundary between this reading of the data and that reading, between this side of the debate and that side. Natural edges are the stuff of life and are everywhere to be found in living tissues, cells, veins, valves, arteries, muscles, bones, and eyeballs, and all through the nervous system, everywhere in the brain . . . and so on, and so on, as we voyage into the interior. Nature equally displays a world of abiotic geological material that "grows," such as crystals.

My argument has been that aesthetic judgment is needed whenever we encounter the cut of transition, whether the edge of the sea, just before chaos creates too much turbulence, or on land, where we build our fences and walls. We should never forget that the most curious character, though essential to the story of Pyramus and Thisbe in *A Midsummer Night's Dream,* is named "Wall." Wall tests our powers of imagination, to be sure—rivaled only by another person in the play, named Moon. With such persons standing before us, the poet seems willing to go almost too far toward catastrophe. In science complexity theory precisely has attempted to address this matter of thinking beyond the boundary line, halfway between pure theory and empirical observation, between natural history in the older sense and current ecology, where scaling and statistical analysis play such a heavy part. Melanie Mitchell's *Complexity: A Guided Tour* has recently given us a rich description of the way networks reveal, in effect, the inner meaning of Euler's Polyhedron Theorem and its relation to ecological concern. She suggests that the dividing edge is a defining attribute of organism writ large, indeed of life's mainly *hidden order,* to use the complexity theorist John Holland's phrase. We are led inevitably to wonder about the many edge-like uses of symbolism, which need defining as our common possession of language, especially when we come to express ideas and beliefs.

Ornamental Edges

Over many years I have studied the role of ornament in allegory, since imagery in allegorical symbolism is strictly ornamental, and this fact determines a truly vast amount of human social life and all its communications on a wide scale. Recently the Princeton University Press publication of a new edition of my 1964 *Allegory: The Theory of a Symbolic Mode* allowed me to append a new Afterword describing "allegory without ideas," allegory violating its hallowed ancient tradition, but the 1964 theory seems to hold up, leaving unchanged an original emphasis on ornament.

The requirement for allegory is a remarkable linguistic fact—in the ancient Greek sense the allegorical icon must be a *kosmos,* which is at once the old name for *a decorative ornamental image and a cosmic order.* Crudely, the cosmology studies an order of images obeying principles of decorum, in that sense ornaments or decorations. Allegories are the image-making genre of ornamental designs and the images in allegory are the ornaments (the *kosmoi)* of any universe. Such expressive means are not always obviously decorations at the edge of a surface, as with costume, although commonly they do provide such edging of things like fabrics. These are familiar devices of framing, which in other terms might be called the bounding of the object. They also may appear in the midst of a larger array, injecting an internal cosmic design, which is perhaps even more familiar from the field of costume and the creation of fabrics. The decoration and décor impress the viewer with a clear index to the decorum of the wearer, which in turn is the result of marking a hierarchic, ritualized "universal" iconic order. Once perceived, this function of the ornamental image begins to take on the proportions of its meaning—order is almost inevitably expressed as an iterated cosmic design of shaping details. Since I have written at length on this subject, I need say only a few more words about it.

The theory of allegory virtually mandates an authoritarian function for language and symbolism. Expression is soon controlled by theme and ideology, even when there is a materialist disbelief abroad that ideologies are "bad." Extreme materialism, or "naturalism" as nineteenth-century authors called it, ends by turning language into a system of rigid signifiers, full of pseudo-religious labels and logos for states of vision that cannot, by definition, be labeled. No wonder then that allegory provides the dialect of the tribe and the lingo of religions. *A fortiori* there is a hardening of symbolic edges, throughout the speech acts of religious propaganda and of scholastic theology (in any "faith" and any historical era). Even in subtle theological argument there arises the opposite of tough-minded thought; there is a web of brittle distinctions, as Francis Bacon noted in his founding philosophy of modern science. While allegory is often said to be a "continued metaphor," it loses metaphorical plasticity quickly, because it obeys a ritual drumming rhythm, denying ambiguity of meaning and hence any semblance of Disparity Action, which we have found to be essential to metaphor. It becomes a kind of "anti-allegory," to use the term richly analyzed by Beda Allemann in a fine essay, "Metaphor and Antimetaphor," on the vexed question of figured language in the modern period. Allemann shows that a figure of thought can undo itself, as he explains how a Kafka parable undoes its own apparently "clear" parabolic message or moral, when its rhythmic order forces the reader to a final moment of uninterpretable naturalism. One can only laugh at the ineffably literal conclusion; one cannot, as with metaphor, proceed further and further into the realm of like and unlike—into the world we actually inhabit. Like the insurance expert he actually was, Kafka forces us through and beyond the domain of countable quantity, beyond the measurable, where even the idea of the incommensurable is beyond measure. He is thus at last the ultimate topological author. His famed humor is exactly of this order—when he explores the madness of pure materialism—and for our time he remains classic, when showing the folly of a rigid intolerance of ambiguity. He sees what is critical about the finally ornamental Great Wall of China, whose laborers never finish their work but believe they are builders.

The hierarchic function of the idea of ornament gets lost in the counting, and the greatest disparity action of all is the juncture between the aesthetic and the cosmic, as if cosmos were the last great metaphor of order. Possibly the cosmos is necessarily a model for metaphor, though I consider it too dedicated to mathematical order, not sufficiently open to its own disparities. Astronomy seems always to suggest an abstract system flawed by its own extravagant violence of motion and collision, and its expansion. Too Wagnerian, one might say—too much the fanatical artisan's total work of art, the ultimate machine.

What is strange about our increasingly technological world is that it is leading into contradictions of emotion—we are at once more and less secure. The obsession with grand unifying theories is unlikely to disappear from human societies, for the reason that obsessions are a governing neurotic pattern—the adaptation to addictive behavior, with its rituals—and by the same token ornaments as hierarchic medallions of faith will not soon disappear, nor should we wish their disappearance too fervently, since after all they imply a passion for a certain degree of order. Their borders and fringes appear to be a necessity for living creatures of all sorts, including the lowly dung beetle Aristophanes found useful for classic comedy. Ornaments may nourish and please the imagination, but they do not contribute a dynamic to the deeper structures of reality. They are somehow extra.

An excessive ornament leads to a trivializing society, which enforces too much busyness. We end up dusting off our annoyances, to quote a friend. As adorning devices our ornaments catch the eye and belong to the larger *cosmic order* of things, but do not drive or build the unifying structure. We want them, like mist on the hilltops crowning the horizon, beckoning to the eye to climb, but they do not establish that horizon. Meanwhile a sentimental view of language depreciates the charm of the ornamental, forgetting that while icons and fringes assume a prior cosmic order, they all too easily limit ambiguity and a sense of difference. A hazard to society lurks here, for icons then become the stuff of sales pitches and fundamentalist literal readings. As with ideologies and allegories, they contribute to the dubious conformism which Harold Rosenberg years ago identified with "the herd of independent minds."

Folded Edges

Not all edges are made to produce colossal collisions between cosmic degrees of difference, however. In his last and in some ways most visionary book, *The Fold: Leibniz and the Baroque,* Gilles Deleuze analyzed the immanent forms of change, such as we find in recent complexity theory and its notion of the Complex Adaptive System. By this I mean a view of change where function and purpose reveal their intentional structure as residing in Nature, as distinct from change imposed by some transcendent Being, a god or a technology treated as *God.* The origin of this contrast goes at least as far back as the basic disagreement between Plato and Aristotle, where the transcendence of the one gives way to the immanence of the other, a dialectical conflict which leaves Aristotle in the camp of scientific investigation, even though from a historical perspective we question this view and we consider modern science to have rejected Aristotle's teleological bent. We find such paradigm shifts within the modern sciences, as Kuhn and others have shown, but it may well be that the early naturalism of Aristotle's *Physics* and *Metaphysics,* and even more graphically the importance he accords to metaphor in the *Poetics,* underscore the broad-based need to understand natural order as an aesthetic phenomenon, where harmony is reflected in various styles. Mathematics as much as philosophy and artwork reflect the desire for such harmony, and in the world of quantum theory the *implicate order* proposed by David Bohm belongs with this search.

In leaning we might say, on Leibniz, it is comfortable to imagine that animals are virtually a combination of mathematics and organic substance. This thought, emerging from topology, supposes a dynamic mode of mathematics, as the British philosopher Steven Connor has observed in regard to the life work of Michel Serres. Serres is a major proponent of Leibnizian thought for our time, and he has always insisted that life-forms are transitional occasions of folding, infolding, and unfolding. A parallel to the Deleuzian fold is to be found also in one of the most fascinating of topological explorations, the Catastrophe Theory of René Thom, where instead of a movement of single invariants or causes of change—let's say, a catastrophic fall in the bathtub, or over a cliff——Thom theorized that seven distinct invariant conditions would cluster to predict a catastrophe. Criticism on two sides, from mathematicians as well as physicists, did not ultimately deter Thom from learning that topological "causes" are mostly a matter of combining families of shaping pressures. This is the bridging phenomenon Euler had earlier discovered, and on which Thom's research focused in a much advanced fashion; as his later career demonstrates, the issues are philosophical as much as physical in all the implications of analyzing stability and change, when the two states are co-present.

Deleuze develops this problem of permanence in change as he considers the artist's stylistic treatment of torsions, a phenomenon of folding that mostly concerns the end of the Renaissance. The baroque sculptural and architectural styles of a Bernini display (and that showy term is not overstated) a strongly biological fascination with growing shapes, where classical austerity is subjected to invariant topological deformations—my previously cited case was Bernini and his sculpting of the translucent laurel leafage of

Daphne's fingers. The boundary between finger and leaf is perfectly ambiguous, not least because Bernini introduces (perforce) a higher ambiguity between imagined living shape and resistant sculpted stone. Carving in marble, Bernini plays on the uncanny contrast between stone and flesh and bone. In similar fashion there is an art of interleaving and exfoliating in the architecture of his contemporary, Francesco Borromini, and others.

Such examples come from the world of art, and they extend outward into what is called Mannerism, but because the actual environing world is mainly a biosphere, the folds of life end finally as a jungle of biological transformations. The ancient Aristotelian biological bias reasserts itself in the Baroque exfoliating style. Baroque visual art projects ecstatic post-classical flamboyance of manner, which indeed historically leads to full-blown Mannerism. This seems odd at first, for it would make the folded forms of the Baroque coterminous with the seventeenth-century rise of modern mechanistic science—a battlegound, almost—but on the other hand this would not prohibit a contrary organicist affiliation, crossing over from the early modern "natural history" of living creatures. This crossover scene in zoology would be enhanced yet further by a slowly mounting (at first alchemical) awareness of the chemistry of life, until during the present period we have advanced microscopic biochemistry. Historically, during the baroque cultural evolution sketched by *The Fold,* the most important imaginative but also intellectual event was the elaboration of baroque artistic forms, and to this day Rome remains perhaps the greatest urban exemplar of baroque sensibility, a swirl always in motion, as Federico Fellini once filmed his *Roma, Roma.* The perfect aesthetic terminus for such a motorcycle ride was indeed St. Peter's; on foot one passes all the way to the columnar canopy or *baldacchino* framing the Pontiff's throne in St. Peter's, where each supporting helical column transforms classic simplicity to a spiraling effect of twisted torsion. To deny torsion to the spiral would be a purblind aesthetic puritanism. Nothing could be more theatrical or more spectacular in execution, since the columns were cast in metal in order to permit a virtuoso treatment. Recalling that topology is simply at first glance the study of bending, stretching, and (as here) twisting whole shapes, we recall that similarly baroque music bends, stretches, and twists melody lines polyphonically so that finally J. S. Bach's well-tempered scale permits a new baroque counterpoint—and "emergent" form is surely what the listener feels, when a Bach fugue unfolds. Counterpoint in more than two, three, four, or more voices simply enriches this process.

Biology remains the ultimate model here, sustained by a deeper music of organic combinations. For argument's sake, consider what we have just described, the virtually mathematical bond between the fold and the baroque artwork. One is furthermore struck by the relation of such ideas to the evolution of writing surfaces, let's say from scrolls to pages, reaching then farther back into historical time. When two surfaces are separated conceptually by a dividing edge, this division can be folded over, making two continuous pieces of writing paper, and with more folds making four or even more pages in a single *signature,* instead of the original two. This permits the folded leaves of a printed book, perhaps the *First Folio* edition of Shakespeare's plays or the much later handset 1855 *Leaves of Grass.* Whitman knew these to be yet one more version of what he called *our old feuillage,* his persistent punning relationship to Nature, and that tells a story to be elaborated elsewhere.

Leafage and foliage are metaphors not only in, but also for the baroque style, as if the page of a book becomes a book only because, not being a scroll, it must employ leafage. Leaves also imply branches, and when Walt Whitman entitled his poems as leafage, he was almost baroque—a thought that might force a rethinking of his whole poetic enterprise, shifting our sense of its lyric chanting rather far from what is too loosely called free verse. Instead we are experiencing a theatrical display, something closer than we might suppose to a sculptor's skill in marble. Like a virtuoso sculptor, the poet stages an artistic illusion of phase-change, and a deeper concept hides here: in logic an *implication* is an enfolded meaning. Logic says that X implies Y, structure follows from vertical and horizontal, as if a neutral entailment has abstractly to occur, perforce, in late Renaissance styles, but the subsequent baroque and mannerist mind asks us to notice something else—the folding, the enfolding, the unfolding that already implies organic process, indeed process in general, and with Whitman something on the order of the logician Paul Grice's "conversational implicature"—a folded sense.

Historically, before Romantic enthusiasm and idealist freedom arrives, the Baroque betrays an underlying conflict, for it clings, like a vine, to the geometric economy of the classic Renaissance forms, and yet it resorts to ornamental cover-ups of this earlier style. The logic of this mixed mentality is complex but strong, for every leaf in the Daphne sculpture is a kind of fold, much as the word "fold" itself implies a prior unfolded object or site, which is now bent into a doubling, or something like a strangely stretched doubling, of its original *gramme* or boundary line. Recall that when with his polyhedral theorem Euler discovered the first edge, he stated that edges are "junctures where 2 faces come together along their sides." Within the concept we are not free to think of the edge only as a singular one-dimensional line; we must also think of its role as creating a multiplying juncture. The edge almost may be said to give meaning and enlarging purpose to the areas of place and space it subtends; it gives birth to the Cartesian *res extensa.*

Redrawing Euler's first discovery, one more time: the line between vertices is Euler's *acies,* but as we now are claiming, when folded it reveals its role as doubling juncture, which in turn implies a principle of organic growth. When trees and plants grow, their leaves unfold along a line marked by a roughly single line. When cells divide, as later was learned, they exfoliate in mostly curvilinear fashion but proceed like leaves along a spine hidden in the leaf structure. When the largest scale of physics parallels this biological processing, it produces the quantum physicist's implicate order. Biological growth requires just such an unfolding "implicature," using Grice's logical term, because life processes in general are examples of foliation, like a good conversation. Lest such visions appear all too organic, we note that the paradoxically postmodern free inventions of Frank Gehry use massive exfoliations to mask and also to support a hidden central structure, as in the Bilbao Guggenheim Museum or the Chandler Auditorium in Los Angeles. Such is the impulse of Gehry's contemporary baroque invention in his unexpected masterpieces of engineering as unfolding. Nor is he alone among recent architectural masters of material leafage. Further along this trail we find now that digital controls make it possible to fabricate all sorts of objects and machines, by esemplastically infolding their parts.

The drift of such connections to the Baroque is that the folded edge is more necessary to life than the metaphysics of a sheet of paper. Our lives altogether unfold as *la vie dans les plis,* to quote the title of a book by Henri Michaux. Deleuze associates this mental play of spatial relations with the Leibnizian concept of a space which at once both encloses and discloses, folds and unfolds, a monadic connection between things that, as it were, need to perceive each other in order, working all together, to constitute a rational universe. Following the insights of Michel Serres's *Hermes,* there is no doubt that the Leibnizian desire for a mathematics of position was baroque in its vision, and if that observation appears rather abstract for everyday aesthetics, we should remember that mathematical thought has become more abstracted than any discipline ever invented, in our time, and yet is designed to be useful thought. Such developments are inherently exploratory, inherently inventing a world of thought experiments. The Leibnizian fold, as Deleuze describes it, is an example of a topological shift from plain lineation to actual folding, a process that permits biological growth. Unless the edge can be folded, it cannot unfold, or we may say, unless the edge of Euler's polyhedral theorem is understood to support a fold-over position, living forms like leaves or animal embryos cannot grow. Ideally the edge is held then to be potentially a double-edged sword. The folding process in turn seems almost to push Euler's polyhedral theorem one step further, as if to a higher dimension of organic change. In any event the folded edge points to an actualizing wealth of implication, a process both materially demonstrable as well as abstractly imaginable, arising from our concentration upon shape and edge. The only way we can imagine turning a line into a fold is to see that any edge may be a blurred, dilated line, a *limen* extended and opened to a place beyond. We are left waiting for a visitor, like Walt Whitman poised always at another threshold—as if simultaneously we needed the symmetry of inside and outside, of home and abroad, yet feared transgressing their edge of radical difference. For reasons yet to be explored, however, it would seem that the Folded Edge is the most important of all types of incision, for every fold carries with it a promise of implication and explication. The cynical realist will of course say that precisely because reason has little to do with successful business or misuse of computers, Natural History calls now for cruise ship ecotourism.

Liminal Edges

In life as in physics, force triggers counterforce, up implies down, left implies right, and so on throughout the vast repertory of symmetrical relations we perceive in both science and art (only God seems free to do whatever he likes, and even he seems occasionally interested in this balancing principle). The concern for symmetry, for example, bilaterally in the plumage of birds and the wings of butterflies, turning to the left, turning to the right, does not imply a fixed or fixating natural phenomenon. It belongs to the domain of process and passage, as with rites of passage in social orders, where symmetry is almost made to be somewhat broken. It is fundamental to all creations of different shapes of things, and more subtly with living bodies, that symmetry appears to be a basic property of the world of art, notoriously when the artwork "breaks frame," as Irving Goffman wrote years ago in *Frame Analysis.* If the mathematician and physicist Hermann Weyl's favored bilateral symmetry seems to guide the forming of artworks as much as living bodies (our hearts being on the left side is a defining exception), and if the symmetries in nature run all the way from subatomic particles to massive crystalline structures, we can hardly avoid wondering how "inside" and "outside" can be symmetrical, except under unreal, ideal conditions. Despite that question, Hermann Weyl and other theoretical scientists have described the omnipresence of symmetry throughout physics. Paul Dirac proposed theoretically that there had to exist a positron, a symmetrical opposite to the already known electron, and sure enough such a positron was soon discovered to exist. But more generally new discoveries could only lead to the question, could there be a fall of parity? Such a condition was indeed discovered by experiment.

In the terms I have adopted for strong imagination, the physicist's fall of parity is a rare subatomic equivalent of the poet's metaphoric disparity action, although as violation of parity Lee and Yang's 1956 theory implies seemingly a quite rare natural event, a sort of conceptual catastrophe. The poet on the other hand is free to continue questioning strict symmetries, which is one reason we need the art of poetry in its most general sense. For the poet the heart is never quite where it should be. Given that such ancient as well as modern frame-breaking is a translation of static symmetry into moving process, liminality has to arise when inside and outside are found in common life to be asymmetrical. The front yard is not the same as the living room, despite changing uses with changing seasons. Every threshold potentially destabilizes boundaries; or rather, a liminal ambiguity of boundary emerges from the capacity of edges to cut through permeable lines dividing two sides, between Euler's topological faces. In this aspect thresholds permit and create contingent folds. Modern freeway crossovers display these with flamboyance.

Boundaries for the astronomer bespeak the "edge of the universe," but space is curved, and that outer edge is a doorstep to a Somewhere Beyond or a Numbered Nowhere that defies imagination, even of the most metaphorically intensified sort. Very large numbers alone can begin to suggest the dimensions involved. Life in a macroscopic common scale of things is perhaps physically simpler, but, as everyone knows about the politics of race, the question of borders may be savagely disputed, endlessly capricious. My concern for the living planetary form, which is ours alone, is to interrogate notions of how to think about a space that is still mostly too large for the individual mind to "see," or comprehend, even on a lengthy visit such as fourscore years and ten. Wars are fought over such vague divisions. We need to reach at least toward a strong notion of the liminal, if we intend to let reality knock on the front door.

Rehearsing once more the topological essentials for polyhedra—vertex, edge, and face—as fundamentals for picturing the globe, we recall that our North and South are flattened and our topography is crammed with irregular bumps and dips, and our edges are equally irregular. Even so, we need always to remember that Earth is still subject to its approximately spherical shape. It has two sides (below and above ground, we could say) and no edges other than natural local deformations or constructed walls. It rotates as a sphere, and its bulging equatorial middle goes all round the globe. In the same way we have seen that while in pure topology it should have no edges at all, in fact edges appear before us everywhere, such that the overall result looks almost like a contradiction and strangely enough to live is virtually to discriminate.

This paradox of edge / no edge is what enforces our need to understand liminality, as locating the scene where edge loses its distinctness of cut. In *Not Exactly,* a book about "reasoning with vague information" and similar topics, the computer scientist Kees van Deemter deals with many different cases of threshold and boundary conditions, where exact cuts are not possible, and indeed van Deemter finds that perfect separations of inside and outside, for example, will permit only approximate clarity. So much, then, for Descartes's "clear and distinct ideas." The edges will only allow for vague incisions. Thresholds are in fact sites of transverse crossing motion, whether for body or mind. Liminal transitions over a threshold typically occur when we experience change from inside to outside, or the opposite. Measurements of a rise or fall above or below a certain threshold are used in all sorts of situations, where the levels of a variable quantity are critical, as for example in medicine. Yet for every medical patient the level of threshold passage is likely to differ from some standard norm—which implies a measure of vagueness.

The idea of time, when crossing over or through a boundary, is equally important in archetypal storytelling—consider Proust's narrative skill at naming such dilated moments. Quest romances frequently use archaic prophetic moments of *mythos,* and in them one remarkably gains insight into a future which is based on perceptions of the present, reflecting the directions hidden in that temporal presence, a turning-time which in turn must be conceived as a gradually dilating passage—"the dawning of an aspect," as Ludwig Wittgenstein called it in his *Philosophical Investigations.* The most taxing aspect is to become aware of our mortality, the victim's death. We can hardly imagine this, but Isaac Babel could, in his story *Crossing into Poland,*which reminds me that a brawny workman in the Sixties told me confidently that "there were no ghetto's here in Odessa."

Tribal or religious conflict may dream of a peaceable resolution by folding and unfolding the source of hostility. Yet the motility of fixed edges may finally, owing to forces of nature, be in fact unattainable or delusional. This leads virtually to a contradiction, when we humans use edges for discriminations for articulating politically a vague mixed-up middle ground, which suggests we need to accept uncertain thresholds.

Liminality teaches a new version of the idea of edge, though proceeding from its most abstract aspect (the ideal topological picture) when it modifies its abstract character, namely through the crossing or bridging of a threshold, which then must be dilated. "How wide can a threshold be?" we ask. On different levels of size the edge becomes the physicists' "event," as if dissolving the hard edges of construction and the changing shores of natural barriers. The doorstep of a house permits movement outside and also inside, inside and also outside; it defines these spaces and markers, as the anthropologist Victor Turner would have wished to say, in processual terms; it treats the development of any event as a series of cuts dividing the line of a passage that moves between antithetical if not symmetrical domains between the inside and the outside. On a very high level of abstraction the topologist Steve Smale in 1958 showed that one could *evert* the sphere, that is, transform its outside to its inside, without folding or cutting into the fundamental shape. The implication might then be that inside and outside are mirror image, equivalent conditions of space. (In a Renaissance study, a 1971 book entitled *The Prophetic Moment,* I showed that in religious or visionary terms the threshold is also the place of prophecy, where past, present, and future commingle, so the past and future collapse into the present. That book was a study of equity, Aristotle's *epieikeia,* in its evolving modern role for law and jurisprudence. Through equity we define and discriminate our liminal differences of judgment.)

In the midst of quarrels, boundaries may be fixed by law, and yet the liminal flux remains operative and brutal conflicts continue, precisely because the exact threshold where the boundary is fixed never loses its uncertainty. To decide is to create doubt, and only those who never look back are free from this plague. There may well be reasonable grounds for rethinking the border, for reconsidering one's property, or the litigant may think he should have asked for more. An estate lawyer deals endlessly with embittered survivors of the dead, whose estate has lost any semblance of reasonable distribution. Political strife caused by redrawing borders is as common as the existence of tribal divisions . . . and so it goes forever. No wonder that Shakespeare's Sonnet 94 includes the lines, "They that have power to hurt, and will do none . . . They rightly do inherit heaven's graces." But his imagined freedom from greed and envy is unmoved and cold, a heavy Stoic price.

Freud's essay "On the Uncanny" is notable for probing rather deeper, showing the sources of the inheritor's rage. In his own Shakespearian account Freud suggests that unexamined impulses fall into disarray, because they revert to early experience or a long lost dreamwork. Uncanny experiences always look backwards through a strange mirror at buried events from our early lives. We are encouraged to investigate what we might we call the dream of the wounding edge. In the same vein we find radical disturbance rhetorically enhanced by unforeseen political opposition, as for example in drawing maps of disputed territories; the opposition intensely generates fear over questions about the supposedly fixed origins of *given* boundaries. The Polish author Jan Kott once told me, from his own experience, that a prison is a place you are prevented from leaving, and though he meant this literally, his work implied that it was also a psychic condition of imprisonment. The border thus belongs to a legacy of time, and the passing of time may never heal an earlier wound.

Boundary crossings are inevitably to some degree memorable, and to convey only the dark comedy of this stressful business, consider the following episode, which illustrates the uncanny strangeness if not open threat of political borders: in the 1960s with two friends I was driving out of Soviet Russia after a stay in Leningrad and Moscow, and we had reached a desolate stretch of narrow road not far from the Rumanian border. Our Citroën came over a gentle rise and in a sharp dip ahead the road looked to be crossing an ominous dark marshland, perhaps forty feet across. It looked like a sinkhole full of filthy wet sawdust, but just before it there was a primitive wooden bar lowered to prevent entry into the marsh. As we approached, up jumped a madman wildly gesturing to us to stop, or so it seemed. He was dancing up and down, and his ragged clothes were the same color as the swamp. Was he a robber out of Lermontov, a starving peasant from a late Chekhov story? He was certainly an anxiety-producing, border-crossing apparition.

Dramatic escape was the obvious choice, so I steered our Citroën right off the road, swerving up a steeply inclined bank. Flying around the trap as we drove away, in the rear-view mirror I watched the mad bandit still dancing up and down with ferocious wild eyes and spastic gestures of rage. Only when we reached the heavily guarded border itself, a scene of further bizarre events, did we learn that the "swamp" was a disinfectant pad designed to prevent cattle from spreading hoof and mouth disease, as they too crossed the border. The madman was as sane as anyone on the planet.

What happened, having reached the actual border crossing, was a perfect coda: An enormous canary yellow Cadillac convertible drove up from the Rumanian side, its passengers a lantern jawed film producer and a bleached blonde actress from Rome. They were on their way to Odessa to make a film about the famous stairs, and a heated, incomprehensible discussion followed between the couple and the guards armed with machine guns; the producer had lost his key to the trunk of the car. That was not allowed, owing to the fear of foreign "literature." A frenzied ensuing distraction, machine guns versus the movies, allowed us to repack our own trunk and disappear to the West. With our last glimpse of the guards we saw them forcing their way into the lid of the trunk, using an electric drill.

Such is the occasional comedy of border crossing. Normally the abstract picture of these situations can be further translated to anxious social and political reality. If under many circumstances we want a folding and unfolding edge, as with organic growth processes, in other cases we ask: how shall we achieve closure and then dis-closure of political borders? Political or tribal boundaries do clearly strive for closures. Yet the availability of fixed edges may finally, owing to some forces of nature, be in fact unattainable or *unnatural.* This situation amounts virtually to a contradiction, when we humans use edges for discriminations for articulating any vague middle ground, and this recalls troubles we face at every scale of human knowledge.

Threshold may be called the ambiguous modification of a modification, where every time one crosses the *limen,* one slightly alters the conditions of crossing. Semiotically, every time a transition across is repeated, one adds a further set of quotation marks for "inside" and "outside," and this grammatical shift evokes a greater indeterminacy of edge. The freedom to pass from inner to outer becomes the issue, not those states in themselves. Finally many crossings produce so many scare-quotes that the original meaning of *limen* disappears, inside and outside merge with each other, and the merged value can hardly be pulled apart—but that is like saying that a metaphor can no longer find its own disparity action. Since Heisenberg discovered the paradoxes of measurement on the quantum level, a similar vagueness about inside and outside has caused much trouble for physicists in their attempts to unify quantum mechanics and relativity theory—and it appears that the difficulty will not go away. In any event it seems true that when edges are treated as thresholds, the imperative to cross continues to stimulate an uncritical impulse to move at any cost, deliberation requiring too much strength of character, much as occurs at threatening national border crossings. Unfortunately, threshold crossings imply the anxiety of making a shift of Leibnizian position, as if to see and live equably with any discrimination were always a vaguely or rationally disturbing adventure, like crossing an ominous political border. Every edge involves what Freud called the "Antithetical Primal Word," simultaneously expressing opposition and connection at the boundary line.

If the thresholds are the primordial form of edge, in one way or another they must expand upon Leonhard Euler's original perception that the cut between vertices (or other similar marked "points") is always an abstract or real division marking and hence creating possibly disputed borders and boundaries. The cut creates but also threatens the continuity of a surface manifold, where values inhere in the exfoliated line, using a cut of only one draconian dimension.

In spite of such difficulties one wishes to say that without liminal divisions, without doors between inside and outside, without edges, there could be no sense of proportion or scale, a topic on which this essay will at last come to rest. Scale makes all the human difference in human affairs, for it measures a proportional balancing. Such balances follow from examining the placement of divisions, discriminations, boundaries, and occasions. What happens when one joins the family of edges? One learns about metaphorical disparities and cultural contradictions, hopefully resolving them into coherent paradoxes, at which moment the edge becomes an instrument of ethical decision.

If this decisive liminal edge is a founding property defining a larger family, then each particular dividing line shares in characteristics of other family members, all of them commingling in the larger topological invariance, which precisely is what defines an ecosystem. Liminality turns edge into a persistent site of change, like a tidal salt marsh blown in the wind, momentarily translating the land into the sea and the sea into the land. We live in such a mingling of time and space, and it is no accident that Shakespeare's last great play, *The Tempest,* is situated in the New World, where time, the *tempus,* actually shapes the boundaries of space. There too time finally brings the promise of reconciliation among warring brothers.

VIII

Shape and the Ethics of Scale

At the origins of topology we found a contrast between form and size, since measurement of magnitudes is by its nature *not* what topology attempts, until for practical purposes made algebraic by Poincaré and others, while scale is always a matter of relative measures. Instead, I believe, our formal theme moves gradually in the direction of Kant's *Third Critique,* where judgments of aesthetic form and expressive quality are the main issue.

Our guiding principle was never more clearly stated than in Johann Listing's 1847 account of the field. He is the first scientist to have employed our modern term, "topology," which he preferred to its earlier Latinate name, *analysis situs.* Continuing in the tradition of Leibniz, Euler, and others, Listing's topology deals not with quantity, but with quality, specifically with the laws of relations between place: *"By topology, we understand then the study of the qualitative features of spatial forms, or the laws of connectivity, of the mutual position and of the order of points, lines, surfaces, bodies as well as their parts and their union, abstracted from their connections with measure or size."* This classic definition names two central properties which had always been in play since Euler's earlier discoveries—(a) topology analyzes position and place, and (b) it rejects any primary concern with different magnitudes and measurable quantities. It manages to sideline the latter by engaging with the former, as Leibniz had wished, so that the question of position could develop into a separate branch of mathematics. We can generalize Johann Listing's succinct definition by emphasizing his point, that topology is a qualitative science, but in an extravagant mood we might go even further, to suggest that topology provides a model for all value-judgments, because it merges the old and the new in a fluent, emergent continuum.

The forming of *situs* leads to a third central principle: (c) objects are seen to be shaped, because their dynamic structure is a *connectedness,* whose properties the topologist studies—thus the sphere can be morphed into a pyramid, pyramid into icosahedron. The connections are a function of formal harmony, entirely abstract, but in both art and the physical sciences, when we ask how the different components and elements link up with each other, for example, biochemically or through combining the colors in a fine painting, we are taking a step toward topological *connectedness.* When topological deformations of site and of body are found to be invariant, despite any apparent surface changes, the topological invariance requires and indeed inspires imaginative continuity. The new idea of an old form creates a new system of linkage in framing familiar data, and in that sense *paradigm shifts* would be topologically new ways of looking at older shapes of data.

Shape and scale are at first glance quite distinct in Justus Listing's definition, although the distinction remains problematic. The differences we recognize readily enough, at least superficially. Visually distinct shapes of things present their immediate appearance of wholeness, contour, texture, color, and overall impression—all of which combine in a more or less constant way. Given such shape-constancy, which rarely if ever has a precise, invariant, single value, we are nonetheless able to identify objects as namable things in the perceptual world. It is as if to recognize something is to perceive its shape, even when, for example, a person's body-form is revealed only by their gait. In forming an abstract argument, equally, we attempt a *conceptual* constancy through correct and consistent usage of the terms indicating, so far as possible, a certain logic. At least with things we see and sense, however, our power to recognize their form is not tied to their size or volume, and hence is not something we are measuring. Shape is not size, but involves a qualitative aesthetic estimate, on which Immanuel Kant focused, although, as he found, he could make only a rough, dichotomous chart, when analyzing such aesthetic qualities in his third and final *Critique of Judgment.*

The scale of things, by contrast, always involves magnitudes and relative amounts, which we might call degrees of quantity. Scale involves quantitative differences located "within a larger scheme of things"—a link between number and size. Historically, human will dictates often that we take more notice of the larger degree of some quantity, 60% rather than 15%, although in a medical or climatological context 15% might be the preferred reading to find. Probabilities have to do with timing, as when we find ourselves regretting that we did not pay more attention to the fuel in the tank before crossing the desert: probably this is inherently a temporal issue, for only time will reveal a probable outcome as accomplished fact. This *should have* is the forgotten voice of attention to scale and might be heard in the voice of engineers who have failed to note similar relative quantities, when building bridges, planning social programs, or testing iced O-rings for NASA.

Conceptual problems arise, as well, when we consider that scale seems to be mixed in a mathematical way with shape. Scale provides a system of ratios which somehow underlies the felt reality of shape, giving every shaped object "more of this" and "less of that" in its overall form. Shape and scale, though distinguishable, finally form a complex hybrid. Measures of scale are expected to bring distinct proportional definitions of difference, but the hybrid would defeat that purpose. Perhaps such hybridity is really a contradiction in dealing with physical magnitudes; perhaps there is a radical conflict between the ratios created though scaling measurements, on the one hand, and the normally approximate, person-by-person variability in aesthetic appearance arising out of changing viewpoints, as when one moves sideways or closer and then further away from a painting in a museum. The work of art in general seems designed to question the basic relations between quality and quantity, and yet topology insists on the human primacy of the qualitative, for a host of reasons, all of which raise difficult if not hopelessly confusing questions about human consciousness: such uncertainties have a role to play in scientific measurement as well, as the design of instruments shows.

What if our ideas of shape and of scale arise from incommensurably different purposes and procedures? The two ideas would then lead us to imagine sharp physical conflict, despite any accommodating mathematics, an opposition which modern art and perhaps art throughout the ages has willingly embraced. As if looking at the world from within, living on a private estate of values, assuming this retired point of view, the artist attempts a rendering of the qualities of things and events; only in a derivative way will the artist render their quantity. Artists, at least in our time, almost always say that the artwork derives purpose and success from personal and cultural taste, not from counting things. The invariant topology inhering in the Polyhedron Theorem allows the artist to modulate around any hybrid permanence in change, meanwhile expressing individual taste, choosing departures from the central invariant which nonetheless do not destroy the persisting inner harmony. Poets have long known this. Regarding a passage in the *Odyssey,* Aristotle remarks that when a part of the story is rankly impossible to believe in, the poet must cover up the lack of coherence with deceptive ornamental language. Continuity in fundamental form is the desired goal. If a Pythagorean number is the archetypal source of harmony in the arts and if, as Walter Pater said, "all art aspires to the condition of music," the poet requires a fluent topological sensibility when translating musical numbers from mystical quantities to felt qualities of life, qualities more "simple, sensuous and passionate" than prose, to quote John Milton. A later author like Henry James would certainly speak of the sensibility enabling such effects, but the same term would fit the gifted scientist. David Bohm refers to "the feel" for scientific relations.

Take an example drawn from mid-twentieth-century science, the DNA molecule. DNA is topologically a conveniently complex instance. In a human DNA molecule there are about 3 billion base pairs orchestrating its double helical structure and according to one estimate more than 200 billion atoms, altogether composing the body of the molecule. Numbers like these do not immediately recall Euclid or the Pythagorean Theorem, for they defy any ordinary notion of form or experienced magnitude when I simply recite the figures, but their scale of reference occurs everywhere in modern analytic science. The DNA structure operates through an intricate, complex, numerous combination and only, it would seem, with much redundancy determines the inherited features of living beings.

Such control of genetic inheritance happens, however, because the molecule has a certain shape, in this case a double helical structure, as Watson and Crick, analyzing especially Rosalind Franklin's and Maurice Wilkins's photographic data, discovered to be the correct reading of the molecular shape. It follows from such cases and indeed throughout biology and everywhere else in nature, that vast numbers of separate atoms and subatomic units, infinitely small bits of matter, form physical groups which for analysis we translate into larger conceptual units. From this aggregate assemblage we get shapes, and without such processes there would be no shape of any kind. If quantity and quality were not thus bonded and bounded by Euler's edges and surfaces, there would be no nature for anyone to call "nature." Nature and its generation, which was originally implied by the Greek word, *phusica,* over time acquired its meaning as abiotic, like atoms, or a biophysical *living* system. Living or dead, the quantity of fractional parts of nature acquires both logic and meaning, when quality and quantity interact at subatomic levels.

For humans to appreciate their biosphere, we have said, they can shrink linear history down to its Vichian cycles, in order to value interactive fractions of natural process. That would at least pull us away from the utopian fiction of perpetually advancing technology that always improves the human lot. If we could somehow *live globally within this shrinkage to a diurnal cycle,* our perspective on linear progress and its unfettered extension would have to modify, as many have suggested. We could bend, twist, and stretch our ideological linearities and cyclicalities, by persuading them to forgo or diminish their unquestioned authority, as it were. Cooperation is the goal here, for extreme Manichaean opposites do not serve us well, and what I am calling the Vichian cycle encourages this cooperation. Somehow we need to debate the way things are and in that way lessen the bad effects of persistent fanatical fragmentation. Looking back to the Romantics, it might inspire us that the philosopher Friedrich Schlegel associated political and cultural freedom with fragmentary thought, which he expressed in aphoristic form. Schlegel admired the wit of Leibnizian intellect, which was prophetic and fragmentary, for such freely discontinuous expression could restrain any futuristic, apocalyptic threats of inevitable global catastrophe or other such dark extremism.

In any event our topological commentary has argued for rejecting such fears of change, in the hope that we might ask how our advanced technology for acquiring knowledge might relate to any dark picture of history, whether linear, cyclical, or otherwise understood. If we worry over the *too much* or *too little,* we must deal with the relation of shape, as in topology, to scale, as in the detailed measurements of the cosmos, including our own planet. To understand the ethics of scale is then to achieve a balance of temperament, to proceed by tempering our demands. In recent years, and I would say, hopefully without presuming, that one great example of this temperament on the relevant global scale—this topological vision, indeed—has been President Barack Obama of the United States. Nothing could be harder to achieve in human affairs, it seems. How then do topological deformations of shape translate into such proportional measurements of relative magnitude, of powers, and what are the methods of this translation? The subject is large and taxing but some outlines may be appropriate to rehearse again: Following Euler's *Polyhedron Theorem,* by combining the Vertices, Edges, and Faces, we always find shapes changing continuously, which means in effect with a certain harmony. Poetry traditionally speaks to the question: for instance, the Epicurean epic of Lucretius, *On the Nature of Things,* while its universe is entirely atomistic, still shows perpetual strings of attachment between all elements, such that a continuity of desire rules in the midst of swerving change, much as we have found in Ovid's *Metamorphoses;* Shakespeare has a line, "Love is not love which alters when it alteration finds," and Friedrich Schlegel's fragments of knowledge and experience are almost the ideal teachers of this sensibility.

Ratios of Order

The Edge or dividing line demarcates a variable training ground for such purposes: it creates fractions. These in turn yield the proportions between adjacent surfaces on either side of the line, somewhat like the areas of adjacent counties comprising a whole nation. Altogether we profit by understanding fractions and their rational number derivatives, including some that are not always ideal; many of these divisions can be made to work in practice. Evolved political systems usually demonstrate proportional power-sharing, as established by the U.S. Constitution for the United States. The arrangement here makes use of boundaries and relative populations within States, and if we could (impossible dream!) get rid of gerrymandering, the system would work perfectly. An ethical principle is involved, for in general an awareness of fractions keeps us from being prejudiced against mere number, and in that way edge trains us to think of an approximating comparative worldview, a measured vision. By tempering the preference for quantity over quality, we fractionally limit an obsession with the mere size of "The Numbers." It may seem utopian, but if money does not gradually subvert socioeconomic benefits, number then enables an even-handed society, for in principle it authorizes a formal language of universally recognized equivalence (the equation, *14 minus 7 = 7,* has a purity of rational sense unknown to ordinary language).

Equitable balance, when treated as a matter of scale and proportion (the "percentage"), shares with topology the goal of variable equilibration and homeomorphic equivalence and beyond these a psychology of what it means, among individuals and between states, to permit a living balance. Though it might sound like moral mysticism, telling is an archaic word for counting, authorizing a famous poem about number, *Among School Children,* when it asks: "How can we *tell* the dancer from the dance?" Telling is rudimentary counting, and for Yeats the enumeration and the shaping finally join into one equilibrist experience. This balance is not always to be achieved by direct assault or theft—consider John Crowe Ransom's passionate, ironic, no less moving poem, *The Equilibrists.*

Topological continuity needs to be fostered. If the living components of ecological harmony—in resources, let us say food to eat, water to drink—those fractions need to be apportioned fairly, while edges of all sorts play a part in this apportioning act. No equable scale can be achieved without first seeing and conceptually separating and then counting (or at least "telling") the separated parts comprising the divided whole. But we know this!

Topologies of Scale

"The world is too much with us," to quote Wordsworth once again. One wonders if there can exist any clear method of lessening the "size" of this world which frightens and is too much with us, since we can know the "too much" only as psychic effect. If that is true, both science and art share a larger combinatorial obligation, in both theory and practice, but surely we need to discover how our scales of measure are continuously deformable. On the simplest level, scale remains the arithmetic ratio of proportional quantities. Using standard units of measure we speak of locating the magnitude of something "on a scale of one to ten." With such linear ratios of relative measurements, we compare one object's size or range with that of another object, and we become skilled at thinking fractionally of the world around us, and perhaps that is an exercise in topological sensibility.

Above all, we use scales in order to measure relative degrees of power, and there we find a conflict with ideas of shape, where there is no clear measure of power. This discrepancy threatens human serenity, surely, since different shapes seem to have greater or lesser "powers" of attraction to animals, and we humans are animals who learn that the energies implied by such attraction are not much like the physical power of a vehicle running in second gear. Psychological power is virtually immeasurable in itself, though we believe we know it when we examine results of animal movements. On the other hand the same is true of all exercises of power—we know their force by noting the movements they produce. We are always in these cases comparing fractions of force and effect, and we do this even in the strange world of quantum mechanics.

A topology of scale might appear a sleight of hand with numbers, as if we could play the proportions of numbered quantities so that their ratios could be understood, in final analysis, to be number-shapes. Then we could topologically estimate their continuity in change, their permanence in change, and to these we could attach values, as we attach values to beautiful shapes.

Perhaps shape is just another case of such cognitive or perceptual comparisons between one proportion and another. We begin to wonder whether shape is another term for a kind of humanized scale, by which we create gestalts of things which we call their shapes, and where, as in topological transformations, all the parts are simply connected.

Somehow the world needs to become more aesthetically sensitive. In that perspective every shape is a statistical curve defining several coordinates, and every shape is a humanly recognizable play of varying proportions among spatial *ratios*—so much length, width, and breadth combined. Architecture provides endless examples, as studied by Alberto Pérez-Gómez. The question arises: can we generate shapes in their direct Gibsonian aspect, by combining and recombining different scales of size among perceived things? In a learned historical account, *Architecture and the Crisis of Modern Science,* Pérez-Gómez ranges through a wide field of *mathesis* during the seventeenth century and later, to show how Galileo's mathematical revolution affected all aspects of architectural theory and practice, especially with theorists such as Claude Perrault. The geometric proportions of a building are its controlling design features, which, thinking of parallels in poetry, as we recall, Sir Philip Sidney felicitously named its *architectonic.* Like builders in stone the sonnet writers wrote their poems in a wide variety of proportional subdivisions, often given names—Petrarchan, Shakespearean, Spenserian, and the like. No wonder music is an archetype for artistic and scientific form, for it seeks proportion. The variations dividing and apportioning the poet's sonnet form turn fourteen lines into a *mathesis,* providing the strength of the structure to stand, to inflect meanings within a powerfully felt shape, like the form of a well-designed building or a well-proportioned animal body when it carries gravitational stresses down from top to toe.

Goethe called architecture "frozen music," which implies a classic principle. Musical *rubato,* chromatic scales, blue notes, and sudden dissonance are well-known devices of bending and stretching melody and harmony, but these expressive bending and stretching techniques assume a stability in the fundamental design of the piece. With the advent of extreme Wagnerian chromaticism it became increasingly difficult for Western music to carve out musical shapes and melodies that actually ended, and with this in mind even Wagner and his successors restrained their chromatic twists and turns. With Arnold Schoenberg, Alban Berg, and the austere Anton Webern there is a later history of atonal attempts to maintain chromatic restraint, but we hear these composers often sounding *Viennese,* as we find egregiously with Richard Strauss. Music is always a clue to cultural style, and in this case the form-breaking chromaticism arises from half-tone exaggerations—a radical change of musical scale from Mozart and Haydn's method. Shuttling back to the seventeenth century, we find that such developments may join art and science together, as if the New Science arose from new thermometers using more and more refined units of measurement, more delicately tuned scales. Galileo had pioneered the study of the ratios of large and small measurable changes in the motion of solid bodies such as cannonballs, while Vesalius in the same era had studied the ratios anatomized as the shapes of organs within living bodies. Perhaps there is a paradox here, but on a macroscopic scale the proportions of internal organs and also external features such as torso to head, ears to nose, ankle to knee, display ratios of form and size leading into a harmony (or disharmony) of shape. Such is the architecture of the living body. Such were the anatomical, embodying concerns of the great Vesalius (1514-1564), who died in the year of Shakespeare's birth. Similarly with bodily movements: the runner changes scales of arm- or leg-motions, when adapting to different proportional sections of a race, while responding to other competitors. As a result, one might speak of the shape of the race, though it keeps changing within limits.

When the English say that someone is "too clever by half," it is the fractional irony that gives us the *measure* but also the ironic shaping of a judgment, and we feel the person's character. Playing on a familiar figure of thought, scale differs from shape in that its variables look out and away from the object, directing thought outward along different dimensions—when comparing standard height to standard width or length, one might think that never the shape, but only the ratio, is the result. But is this so? It might as well be argued that analysis of a site or a character or a situation is to accord a contour to those elements. The scientist might naturally observe that science gives the observation its true theoretical frame and then its status as factual, whereas shape does not look outward beyond its own enclosure of a site. If so, we might say that scale is the outward, comparative, quantitative, relational—even universalizing—aspect of qualitative shape, while within large populations assigned *valuations* are bound to differ. It is noteworthy that the term "value" differs in meaning as much from moment to moment, as there are fields of human activity and thought.

In searching for values, the aim of statistical analysis and reading of objects *en masse* is to get away from the massed jumble, while a wanderer, dreaming almost, studies a cloud and sees an independent closed form, the outline or boundary of the cloud often appears vaguely dispersed at the edges. Yet the body of the cloud is a place within, for the cloud is independent and while indeterminate, it is still a Oneness, a distinct component of the biosphere at its atmospheric level. Though knowing and seeing all the shades and streaks within the cloud, a landscape artist like Constable will nonetheless make that Oneness virtually the archetype of natural form, the final cause of his paintings *en plein air.* Our Leibnizian visionary, Michel Serres, analyzing the mind of technology in the nineteenth century, would be more interested in the steam engines and the wild vaporous torments of Turner, who thus was able to bring clouds down to our biosphere, our Earth in every sense. Turner belonged to a Victorian age of measuring scales of power, seeking to evaluate external forces, operating frequently in a random mass of discontinuous steps. He was not far from the period when Einstein, following other physicists, began to study the Brownian Motion of randomly colliding particles, nor from Karl Pearson's first use of the term, "random walk." Random such arrays and paths may be, but when we evaluate them, we find the shape of Brownian complexity meaningful in relation to externally measured ratios of mass and energy, and hence of power.

Illusion surrounds all evocations of power, as when today people speak of empowering the individual, and this illusion arises in connection with the idea of scale. For scale remains entirely relative, without any identity other than comparative percentages for a variable; it works within the variable as such and in itself remains a merely useful if strict quantitative comparison, a sort of metaphor without substance, an index to the odds of something happening.

Topologically, however, something else is happening, which we may call the appreciation of quality; this may be identified with the connected, continuous flow of transformation, which occurs when we relate objects to their positions and to their existence independent of their size, and as a result allow shape and form to be an ethical influence in our world, without regard to serving any materialist Dickensian Gradgrind function. In that way shape becomes an exemplary model for various harmony, a flowchart beyond price, and this freedom is an entertainment of fractions and fractals and other Hegelian "concrete universals." Quantity would then serve judgments of quality, as proportionate parts of a larger combination, which is exactly where the artwork focuses its labors. Quality is a mode of belonging, perhaps only expressible in ordinary language, making no pretense to universal mathematical neutrality, and as such we recognize quality in art to be only an imagined harmony. A sense of quality allows the mind to break free of rigid ritualistic constraints, to improvise, to think like a natural athlete, almost in that way to become theoretical, by redefining proportions and scales as the first mathematical footsteps toward an appreciation of shape, most especially the lasting invariance of changing shapes discovered in topology.

In our present historical period we think statistically about almost all global judgments, and the global picture remains in many ways a consequence of our electronic networking and information processing. We are now faced with the problem of enlisting the individual to participate in the network, as if each of us were a Hamlet or Ophelia trying to assert their individual characters, yet hoping to be connected in a courtly network. Why is the clown a gravedigger who digs up the skull of the court jester? "Alas, poor Yorick." Will Polonius win? Will Claudius the murderous usurper triumph? What of the Queen, in all her natural complexity? We think we are the spectators, but we are actors in the drama, and this paradoxical drama, this situation, besets us all, including *neutral observers,* and so I argue that we need an opposing vector of topological imagination where shape-shifting always holds to an invariance, while the invariance encourages public character and self-discovery—the quest in opposite directions by itself gives purpose.

Fractions of Purpose

Throughout this essay I have been moving back and forth between Arts and Sciences and also personal memory, by suggesting some analogy between them, which even though Science and Art may not always be happy to see it—a shared fragility of human purpose. We need some specific ground for thought, especially when the questions are very large ones and always changing their shape, and we keep looking for that ground when thinking on a global scale. As a beginning we might ask which of the two approaches, the topological or the measurable, the shaping or the counting, would be of greater help to our world today, as technology and populations grow at an accelerating rate? Which will guide us more usefully but also with greater fairness? Which will reveal our worldwide direction of change with greater clarity? Yet we have seen that the two approaches need each other. These constitute a single complex question: how shall we proceed in deciding on the better choices of position on an electronic map conjuring the same size as the globe it maps? Geographers know that we cannot simultaneously project the surface of a sphere onto the flat, keeping both size *and* shape—we must choose which of the two we wish to map; we cannot have both. In every mapping, however, the game *begins* on the plane of topology, where we discover a scarce visible abstract mapping of space itself, the mathematician's vision our great poet, Wallace Stevens, called "The Motive for Metaphor." This poem the late Northrop Frye made central to his lectures on *The Educated Imagination,* and there the poet chants of "the half colors of quarter things," the obscurities almost beyond expression—the motivation to participate in the powers of ordinary language and its figurations.

Metaphor, finally, is a semantic device for showing the relation of scale to shape. Such is the meaning of the *disparity action,* a break in identity necessarily comprising the lesser quantity within the fraction. Otherwise similarities will disappear, and chaos will ensue. Metaphor, we notice, breaks frame, so that perhaps there is no final goal to be reached by linear advance.

There was, of course, also a suggestion that we might find a final question about shape and scale. Unfortunately there is no final anything, and the monstrous idea of finality is a mirage in an unexamined desert, like ponderously believing "in the final analysis," as if that phrase should be spoken in a serious and appropriately modest tone. In human affairs, as long as we are conscious, the only answers are provisional, and that is precisely what Leonhard Euler must have understood, when he found that the Königsberg puzzle could not be resolved by common sense or arithmetic and counting, and later when he saw that the five solid Platonic forms could be mysteriously changing and invariant at the same time. In the upshot he redefined place, to remove it from the category of fixed singularity (*this* bridge, *that* island) and instead rethought it as position in a complex network of possible choices. The choices had to be seen as a sequence of path-decisions, where the city itself remained the same situation. Of this site the *fractions* or significant parts of the whole system are created by lines or links (the *Edges*), which are thrust forward as bridges between distinct urban areas (formally the topological *Vertices*). In topological terms this complex arrangement permits only certain combinations of crossings, and ultimately Königsberg became a combinatorial puzzle. Our modern legacy from Euler's mental graph of this setup is that its bridges and vertices become the links and nodes of digital technology and its combinatorial algorithms. In a primordial sense Euler had discovered the modern network, by imagining a higher order of fractional design.

Evidently imagination plays a main role in such visions, a role compounded of two interests, shape and scale. Coleridge and Schelling before him had pictured a ladder or scale of developing vision, starting below from fanciful agglutinations of separate things (the Puffs, Powders, Patches, Bibles, and *Billets Doux* in Alexander Pope's *Rape of the Lock*), then mounting to a unifying *esemplastic* vision where many different levels of reality merge together. That resembles a movement from simple arithmetic to higher mathematics, and it pictures the obscure way scale and shape interact. Granted, it may appear that I have been trying to fabricate shapes by using relations of scale, and in a way that is true, for every shape may be seen to have discernible proportions, such as the feet versus the legs of a human body, or the ends and sides of a rectangular box. A cartoon image of a robot will resemble a real human body, simply by drawing a large boxy shape for the torso and a smaller one for the head, and with similar proportional games one can make a cartoon of a fancy modern TV set that is almost robotic. The proportions do convey the typical outlines of a distinct and simple shape, and yet the proportional approach always leaves the construction of shape in an unduly open and uncontrollably relativistic condition. We are left thinking, "Yes, we have this cartoon shape, but it is distinct from what?" Poincaré once asked if we have any idea of what a point in space might actually *be*. If that is an unresolved question then we also ask, "The scale from which it is derived comes from where?" Perhaps it issues from a model drawn from thin air or geometric thought. Perhaps the point and the scale are unknowable, imagined singularities. We are left asking where the initial choice of scale comes from, in the first place, abstractly or realistically it matters not.

Owing largely to the noise of instant communication, background collapses into foreground, and our minds today are driven to an overwhelming yet inexplicably vague anesthesia, caused by an almost robotic futurism underwritten by too much information to process. We may experience the panic of the *too much*. The poet Wordsworth was not panicked, but everywhere he saw the early stages of the disease, nor could he or his contemporaries brake its course, which like the irresistible ocean in Shakespeare's *Coriolanus* "will on the way it takes, cracking ten thousand curbs . . . asunder . . ." (Act I, Sc. I, lines 71–73), —for we are creatures of infinite desire. The danger of linear progress is that it brings our dreams too close to the surface. We lose depth and we lose time when we are directly identified with astral bodies sailing in deep space, where vast magnitude inspires no emotions but uncontrollable awe or unappeasable hunger for more. Abstract the numbers may be, but they are always beckoning beyond us, like an unaffordable marriage with the stars, or so their fragmented foreground makes its promise.

If the world is too much with us, that means that the scale of things in every sense is too much to handle and we are ourselves expected to be out of scale for reasonable human abilities to deal with this global *acceleration,* as Bassler calls it in *The Pace of Modernity.* On his view we can imagine a hard quest: What I am calling the topology of scale aspires to map the way numbers and their scales mimic permanence in change, as if the number line were an infinite wheel bringing continuity, a continuous flow, to the changes we impose on our world. Topology of scale would try to show how knowledge and information in the abstract are shaped, how they are not yet one further neutral cluster of numbers, however scaled.

Surely there is a mental and real conflict between the measures of things and their shapes. We are already on that road when we think that statistics are finally the only sound way to understand our most advanced knowledge. Are these measures a secret sales pitch? We are bent on inventing obsolescence, however it is disguised by comparisons of different scales, whereas a considered topology of scale would indicate that shape is always a kind of governing, balancing motion in time, composed (if we can use that word) of purposive fractions expressing the meanings of time passing.

Shape then is *always* modulating *a permanence in change* and is what we mean by that phrase. Twentieth-century science began with the mingling of time with space, and so it has always been with the arts. The most moving instances of questing this Lacanian inmixing are perhaps to be found in literature, in the recognition scenes of drama, or in the novel, which virtually begins with Cervantes, whose Don Quixote discovers his own and Sancho Panza's lifetime, but we have many other examples of shape which test our imaginative and technical powers, these tests being musical in origin—consider Balanchine, the always unexpected, classically trained choreographer, or Cézanne the painter when representing shape, amidst the artist's frightening uncertainty, so powerfully rendered in a famed essay by Merleau-Ponty, *Cézanne's Doubt.* With these masters there is little inclination, perhaps there is no room, to count or "realistically" measure the sizes of objects, since the dancer's body, moving, creates the size required for the choreography. Embodiment organically seems to merge the distinct attributes of quality and quantity—an imaginative psychosomatic union.

When we map the curved surface of Earth onto flat pieces of paper, we demonstrate the abstract principles of the topologist's similarly moving "shape of space." Why then should we not explore the shape of scales? It is possible that the mathematician's "scale invariance" permits topological transformations that make sense, however strangely, for objects then do not change, even when energies vary, and we end up with "dilatations" at every moment of existence. Altogether we are left with the sense that topology maps a real aspect of our otherwise unreal human world, in that what we take to be real and full is only what our minds are able to "construct," which encourages us to return to the question of the ethical value of considering proportional and spatial relations, dimensions of reality which in the same way we also "construct" according to beliefs in different cultures. Finally all measures are involved with emotive value, but although they imply communication, they point to *felt* shapes, whose nonanalytic recognition is what the physicist Michael Polanyi named "tacit knowledge." The moment we share our conscious thoughts with another, we allow others to sense our feelings, even if only in part. Because things perpetually change, there must be permanence in change, and our quest for knowledge turns on the idea that time is passing, even as we try to fix the forms or shapes of the events—Whitehead called these the "occasions"—of existence. This occasional aspect of life suggests that we are irrevocably isolated creatures, wrecked on the island of time, frightened of the purity and danger of our achievements. The aesthetic isolation or individual separation of every shape, an infinitely separate identity quite beyond number, is where we are finally forced to settle, and that aloneness in space is where we make landfall. Let us hope that in the pursuit of a well-formed curve of continuing biotic existence on a human scale, we are not undone by our obsession with the numbers. We have much of the skill we need to advance our calculating prowess, but what we lack is the judgment to understand how and for what purposes we should employ that skill. "Of course," the pragmatical cynic will say; but then he may have given up on caring and understanding altogether.

The issue of scale determines the future of life on our planet, and it follows that an ethics of scale is the critical desideratum. Only in its light will we avoid disastrous cynicism and insatiable hunger for power. Yet only a hopeless optimist will find this ethical ideal to be easy and cheerful, for obvious material and political reasons—the famous Renaissance "reasons of state." Today's serious ecologists find little reason for optimistic (or utterly catastrophic) visions of the future, since our species seems unable to evade the lust for power, a momentary dream, to be sure. We are finally forced to confront the question, to what extent does survival drive us into greater and greater isolation from each other, as tribes, as individuals, sliding always backwards to prehistoric tribal man? We need better edges for coherence.

IX

"No Man Is an Island"

Amidst an extended encounter with so many shapes of disparity, so many fractures of continuity, so many distinguishing edges, I have been seeking examples of the connected topological imagination. One example might be the imagery surrounding our idea of inhabited space, our sense of a home or neighborhood to which we are peculiarly attached and in which we find many resonances of attachment. Space and place thus inhabited may change their outlines, and yet remain "the same place," as we have supposed so many times, and we therefore question this sameness as yet another topological "permanence in change." Meanwhile any clear doubting reaction shows there can be no conclusion to any essay, like this one, whose exploratory, wondering, unfolding trial continues in the mind.

My provisional ending then is an "emergent occasion," adopting the phrase of John Donne, a great poet and preacher of the English Renaissance. Our emerging human question looks to the horizon, reconciling the human species to the limits of our sphere, our cosmic home, for it and we are fundamentally *alone in the universe,* to use John Gribbin's haunting phrase. Perhaps one answer might come to us from an unusual event occurring in a long-ago time, which passed just yesterday.

In his *Devotions upon Emergent Occasions, and severall steps in my Sickness,* the Reverend Doctor Donne, whose youthful poems were well known metaphysical displays of wit, like his later sermons as Dean of Saint Paul's Cathedral in London, serenely took some most unusual notes on the progress of a deadly illness, his own in fact. In December 1623 he fell victim to a contagion whose pulsing symptoms rapidly progressed, leaving him without much hope, until without warning he called for pen and paper, to create a complex, documentary, *simultaneous* account, resulting in a ritual expression of relativity. In the course of a spiritual diagnosis he pretends to be writing the simultaneous account of the illness even as it progresses. The contagion had already killed many Londoners, and perhaps in a ministering spirit Donne's account was published in early 1624, astonishingly soon after his recovery. In his later years Donne had become famous for his pulpit rhetoric, and here again he spoke of sanity and affliction in learned, exegetical, yet dramatic terms. He even brought the helpless medical doctors into his account. The crisis passed and he could pray with gratitude for a miraculous recovery.

Of much troubled and suffering Catholic descent, after 1615 he was yet a Protestant member of the Church of England, and throughout his strange metaphysical narrative he continued to invoke a mystical union with other members of a single composition, the "one volume," as he calls the "catholic universal church," a belief and language still persisting in Anglican services.

Sometime before the happy medical outcome, while the victim's notes continue, a funeral bell tolls mournfully off in the distance. The bell tolls for the passing of an unknown stranger, somewhere in the distance, and Donne memorably writes of this unknown: "Perchance he for whom the bell tolls may be so ill that he knows not it tolls for him," but he continues the meditation in a vein of ironic geography, as if suddenly knowing the ultimate solitude and anonymity of death. The words alone *place* the moment of passing from life to death.

> No man is an island, entire of itself; every man is a piece of the continent, a part of the main; if a clod be washed away by the sea, Europe is the less, as well as if a promontory were, as well as if a manor of thy friend's or of thine own were; any man's death diminishes me, because I am involved in mankind, and therefore never send to know for whom the bell tolls; it tolls for thee.

This passage is a musical monody, so exquisitely expressive and controlled are its unified thoughts, words, and rhythm: we must, however, interpret Donne on more than one level. He writes to an owner of manors, yet he prophesies a mystical Christian vision of one single person belonging to a transcendent multiplicitous unity, whose symbolic intention would be recognized by followers of many religions, while simultaneously, as with other sublime utterances, there is a wealth of other visions sustaining its metaphors.

In the first place, Donne's vision of a paradoxically continuous union between separate domains is entirely imaginative; without the elevating power of imagination he could not have discovered the connection between the resonating sound of the bell and the union of earthly masses bridging an empty oceanic space. No better instance of the Coleridgean *esemplastic* power of imagination could be given, indicating a sublime ascent higher in a Longinean sense above the Coleridgean level called "Fancy," raising it to a real yet transcendent level of full Imagination.

He is not, like an engineer, thinking of seven built bridges but as theorist of position—a vision anticipating Newton and Blake, as if inspired by Plato's *Timaeus.* The negative idea that "no man is an island, entire of itself" is imagined as simultaneously implying a containing, Platonic, Timaean *chora*—the "continent," he calls it—such that every island is also an immersible "piece" of the mainland mass. This is a topological metaphor, through and through, even to the point of stressing its inner disparity action, for it is clear that every man is nonetheless also an island. Island and continent somehow merge or join.

In the second place, the passage depends heavily on the basic topological contrast between shape and scale, for in his isolation each individual person is a poor weak human body and soul seeking to be transformed, without essential loss, when it joins the larger "body." This larger continent is imagined to incorporate the single personhood into a vaster scale of being.

In the third place, the topological understanding of shape and scale has a linguistic aspect: the figurations expressing its thought in subtle metaphorical play, as with words like "main" and "manor," or on the many senses of *island* and insularity which Donne, one of the translators of the King James Bible, would conceive as emerging senses or meanings. Following rhetorical tradition and expanding on metaphor proper, the chief figurative method here is a kind of synecdoche (A participating in B) and also metonymy (A and B having a named locality), where the similarities between things are tied to their being close to each other, while positional adjacency becomes a mode of belonging to a set. This language points to actual, troublesome, and extremely difficult social and political conditions.

In the fourth place, scientifically we need an even broader interpretive view which aligns implication, *infolding,* with the topological transformations David Bohm named the *implicate order,* where art, science, and perhaps we should say ethical values combine. After the Copernican Revolution and all its immediate upheavals, there would be no smooth road for the advancement of science, certainly not before Newton's *Principia* was published in 1687. Such beginnings were for Donne an intense experience, which troubled his unquestioning faith; in the words of one of his *Anniversaries* they brought "all in doubt," while a Baconian empiricism also implied a fresh conception of human destiny. Donne seems instinctively to have looked far beyond the beginnings of modern thought, making a hopeful but sometimes dark poem of the probable future. Europe's hunger-driven Thirty Years War made the darkness all too visible.

Among twentieth-century works, notably *Wholeness and the Implicate Order,* David Bohm went beyond his own field of quantum theory, to explicate the implicit—the infolding that allows physical nature and human nature to cohere into a larger order. Bohm as quantum physicist thinks in terms of combinatorial physical coherence, where change is a constant condition. Coherence rather than *logical* consistency is the global issue, for example in the topological leanings of Gilles Deleuze, which appear centrally in a book on Leibniz and the baroque, *The Fold,* but also in later meditations, *Pure Immanence* (2001) and the first essay in *Desert Islands and Other Texts* (2003). Deleuze writing on immanence explicates the phrase, "involved in mankind"; we are reminded that the orthodox Christian Donne breathes also an air of secular materialism combined with "natural religion," here in the *Devotion* no less than elsewhere. Like Sir Thomas Browne in the 1643 *Religio Medici (The Religion of a Doctor),* Donne the spiritual doctor reaches ahead to a Romantic imagining, where objects may fuse with transcendent realities, whose mixture shows a poet standing at the threshold of our modern world. He craves an elemental, implicate, folded topological harmony linking the isolated with the widely extended, a mysterious yet material edging and bridging, foretold in lyrics like *The Relic* or the *Valediction Forbidding Mourning.* For our time, in a not so dissimilar way, Bohm's implicate order provides just such an imaginative construct.

The four intersecting ways of reading do not exhaust the possibilities, which might be approached through the open method of William Empson's *Seven Types of Ambiguity* and *Structure of Complex Words,* though they make a powerful beginning. Above all, the four approaches point to a typical seventeenth-century manner of crossing back and forth between the sacred and the secular. Looking far ahead into a future of our present needs, Donne's vision of a modern world anticipates our own concern for a wider ecological science, whose metaphors of the separate ecosystems are joined to metaphors of islands as parts of an intelligible manifold, to use the topologist's term. For a new age of art and science—and religion—Donne brings a sublime metaphysics down to earth.

If islands and continents involve theories of the world, seen from different philosophical and scientific angles, represented in language in complex ways, they also appear in a drama of immemorial storytelling. They are demonstrations of what William James memorably called "the dramatic temperament of nature." Because logos and mythos complement each other, it is no wonder that the facts and figurations of islands fascinated the poets, painters, and travel writers of Donne's epoch. The mythology in part descends from the biblical Garden of Eden, now darkly reborn in *The Lord of the Flies,* an island or isolating motif frequent in much of William Golding's other fiction. Such tales, both light and dark, carry with them a cargo of beliefs in innocent and not so innocent sexuality—the serendipitous benefits of empire, greed, and adventure. Imperial hunting for natural resources is not news, of course, but what may get academically buried here, unfortunately, is the global aspect of empire, which virtually defines cultural practices. Geographically, the islands are only one significant global site, but they constitute a very special geographic phenomenon, whether imperial or not, and they have their mythology. Ancient Greek and Roman legends had included that of the Fortunate Isles, located vaguely somewhere far away in the Atlantic, while Pre-Socratic philosophers flourished in a culture of actual Mediterranean islands. In their Ionian wake we recall a typical echo of Homer—the Island of Venus in the Portuguese epic,

The Lusiads, written by the extravagantly adventurous Luis de Camões to celebrate the famed earlier explorer, Vasco da Gama, who had sailed eastwards to discover the Indies. Italian romantic epics, notably the *Orlando Furioso,* imagine space travel and see islands of love and war everywhere, to be followed by the great English poet, Edmund Spenser. We recall that Shakespeare set the scene of *The Tempest* on a mysterious Bermudan island; in another play, *Richard II,* he called his England "this sceptered isle," a "demi-paradise," / "This precious stone set in the silver sea, / Which serves it in the office of a wall / Or as a moat defensive to a house."

Perhaps the most intricate and philosophically interesting use of an island kingdom belongs to the early seventeenth century, when Miguel de Cervantes parodied the idea of absolute rule, with the misguided Sancho Panza obsessively wanting to govern an island in Part Two of *Don Quixote,* the first modern novel. An island, ruled, was in principle a body of land one could truly *possess.* Similarly, acquisitive explorers searched for the Spice Islands, almost as if natural wealth had always to be a separate treasure taken from Earth and its cultivated progeny, but essentially whatever is found and taken as a prize.

If we ask why islands are so special in this context of possessive rule, our history must take a more scientific turn. The account book is almost too rich with detail, so that among evolutionary scientists we need only mention the names of Charles Darwin and the Galapagos Islands. In more recent times we have the *island biogeography* of Robert H. MacArthur and E. O. Wilson. This now classic 1960s ecological program, which owes much to Evelyn Hutchinson's example, was pioneered by MacArthur and his colleague, when they analyzed the transactional dynamics of movements and shifting distributions among species populating archipelagos and other clusters of islands, such as the Galapagos. Regarding my own much more generalized topological context, *island biogeography* brings out the fundamental barrier that confronts the wider question of terrestrial macroscopic place, a scandal at times occurring too fast for us to handle, when, using my analogy, this idea of islands allows us to examine natural habitats disturbed by our human expansions, while human habitations may be destroyed in the same predatory way. For example, an island is a "natural" phenomenon, the scene of MacArthur's "natural experiments," but we fence off our isolating *reserves* for wild creatures, which are wrongly confining spaces for their survival, and with somewhat less than the best will in the world we give them "corridors" or Königsbergian bridges for passage from this isolated place to another. This does not work at all well. Beyond these changes, we have continued to see frightening displacements, by brutal force, of whole populations of humans.

Precisely because island or "insular" biogeography can be extended to include all sorts of insularity, the original equilibrium model remains an important testing ground or imaginative schema for thinking about the broader life process. Islands are not only isolators and thus protectors in nature, if the island is impossible of access, but that isolation frequently though not always works to the advantage of the protected species. At once we notice that cities and towns and villages are also islands—we are in a turmoil of conceptual expansion.

By analogy it seems unavoidable for us any longer to deny that Donne's *Devotion* and the recent island biogeography speak with equivalent tones to our human species and its movements across populated insular spaces, with similar and similarly disturbing dynamics. On the side of both art and science, therefore, the preoccupation with wild nature's magic and its picturesque or sublime insular separations from cities and towns dangerously drop down into our dreams, where we associate the insular archetype with a naïvely safe discovery of new kinds of life, and every such discovery feels like finding *Treasure Island.* We continue to forget the numbers of species we are destroying every day of the year. Everywhere we look, we find that a changing sanitized world leaves a trail of trash and almost living matter.

Here too a metaphoric disparity needs to be examined, for islands are I. A. Richards's "unlikenesses" within in a wider geographic manifold. In the broadest "catholic" sense Donne had said that the species *mankind* is both isolated and not isolated, and this in a mystical sense, which we should not violate. Spiritually we should not be forever segregated. One wants to claim that this view is neither secular nor scientific, but it may lead in a secular and scientific direction, and in an older sense a religious attitude toward life, at least at this present moment of the twenty-first century. If the words *island* and *isolated* (from the Latinate *isola*) have a common meaning, we perhaps should combine idea and fact, theory and observation—our secular version of the spirit and the letter? Let the following stand for endless parallel situations—suppose that our global question is the following, "Who is isolated and exactly how, from whom, especially in relation to our larger environment?" Gordon Baskerville, the prominent Canadian ecologist, insisted always on a distinction between local pragmatic management of forests and equally on the detached theorizing of ecological model builders. In history this is a familiar conflict, but Baskerville points out that we have the twin problem of connecting the scale and connecting the shape, when we analyze global ecology. He wants a functional dynamic built into the theory, and thus an infolded and yet edged analysis that can be actually tested by managers and developed by theorists. The difficulty inheres in our needing to deal precisely with larger unities of larger groups, understanding both forest and trees simultaneously.

Geographically and also as ideal mathematical objects—essentially we might say—all islands are *topological features of the surface manifold of the Earth.* We commonly use the word *topographical* for such unexpected features. They spring up, mostly volcanic, rising from a lower lithosphere, or if in a continental flat area the surrounding solid ground falls, revealing the isolation of an island shape. In massive cases, like the Grand Canyon, there is an immense, very slowly moving interplay between rise and fall of shapes that end up looking like an upside-down mountain range. Thus materially the island is striking because its concretely visible shape is equivalent to the idea of stretching and bending and twisting the topological surface. Earlier travelers were aware, for example, that the British Isles belonged with a group of "continental" islands that somehow had fallen away from a larger continental shelf. Later voyagers could explore very different scenes of sharply separated volcanic islands such as Napoleon's last prison, St. Helena, or Conrad's fictional Samburan in the novel *Victory,* not to mention actual strange coralline formations of atolls, all of them edged by bodies of water. The shapes and origins of islands display many different geological forms, but lying beyond the ocean main their horizon is not necessarily the sea, since geologically a sudden granite outcropping might look all around toward horizons of flat earth stretching far away, in the midst of Russia or Australia. By yet further extension, in some archetypal sense every island is *stark isolation, a deserted place*—the scene Daniel Defoe developed in his politically prophetic novel, *Robinson Crusoe,* which with its sequel, *The Farther Adventures of Robinson Crusoe,* gives us no doubt the most important analytic treatment of ideological isolation. To comprehend such translations of the physical into the mental, we need always to recall our uses of metaphors and their disparities, and the need to appreciate their projective power.

An island in the sea cries out for its map, its promise of treasure in war and peace being a kind of encapsulated gain, an archetype represented by old coins. As long as one does not have to actually live in an island's isolated environment, always the same, one is delighted to think of its magic. Yet there is an odd way in which any map or chart isolates the manifold surface it is mapping, as if to insist on the act of incision into the real. But in the context of "dramatic analogy," as I have called our experiential rather than logical truths, the map and the place forge a strange contact, an imagined physical *touch*.

As children, I remember, all was a world of changing and moving sights. We were thrilled that the notorious Captain Kidd had buried his treasure on Gardiner's Island, immediately across the water from our old Long Island house called "Fireplace." We never asked where the treasure came from—after all, in a Three Mile Harbor boatyard we gazed at rotting high-speed motorboats and longed to become *Rumrunners,* whoever they were—but when Sarah Diodati Gardiner, who then owned Gardiner's Island, showed us a small piece of gold cloth from the treasure, we were disappointed it was not grand like the coins of our imagination. In summer we could sail across the bay, over to Shelter Island, passing the forbidden Plum Island—why was it "forbidden"?—and then returning round past the old whaling town of Sag Harbor—it was before the developers had displaced the potato farms and pine stands—and all such experience was tied to the intricate magic of island coastlines, where the direction of the wind and depth of water effectively change moment by moment. Was there a more important kind of island life to be understood? Perhaps life itself is a figure for an island lost for lack of an ocean, let us call it the Absolute Island. If so, among such ideal forms this Absolute Island misleads us, when Donne's *Devotion* promises an impossible unity of all mankind. Or perhaps not!

If each island can become the mainland for an inhabitant, and the continental converse is also true, there follows one of the great problems of geopolitical power: somehow we have to lessen the world-ignorance of what usually goes by the name of "isolationism." For geographic reasons many fortunate but troubled nations have promoted their own isolationism, as Roosevelt knew in relation to declaring war on the Nazis. Established bourgeois comfort virtually guarantees the isolationist attitude, so we see how ambiguous are the ideas surrounding the island as metaphor. In an advanced society where as long as money can buy it, like the lady's water supply, our supermarkets are displaced occasions of a comfortable global inequity, in which we can hardly fail to participate, even as we decry it.

The grim, dark side of religious wars must be reckoned in the present scale of power plays and random violence. Our planet appears so close to a chaos pretending to be organized, a new fanaticism, that we readily question John Donne for his eirenic mystical hopes. An establishment figure, in spite of the tragic sectarian deaths and sufferings in his own family's history, he could hardly have foreseen the horrors of the Thirty Years War, though they had already begun when he wrote his *Devotions.* What already had changed was the accelerating military expansion of industrial progress during the seventeenth century, as if the whole world were headed directly toward Stephen Crane's *Red Badge of Courage* and Cormac McCarthy's ominous tale, *Blood Meridian or the Evening Redness in the West* and beyond these imaginative works the litany of demonstrations of the horrors of war. Journalism does not help us here, its trade being repeated mindless shock and sensation.

Informed persons are only too aware of this likelihood continuing unabated in our time, knowing that Peter Sloterdijk is right to insist this involves the powers of cynical reason. It is as if the empty miles of forcibly emptied spaces were pretending to be real human extensions, inhabited places. If more and more dangerous religious wars continue, we shall hardly know what is island and what is continent. The air itself is no longer safe for travel. How then shall we give sensible realistic meaning, in any way, to Donne's *parts of the* [Spanish] *main* which were strictly speaking stretches of Caribbean ocean, or shall we find wholeness in the world's terrestrial mainlands? Boundaries may cease to have environmental meaning, even when decided by surrounding seas or geographic lines, certainly when reflecting warring political histories, driven by resource claims. In the end what then will be the meaning of human place, as the philosopher Jeff Malpas asks? We are not wrong to ask why islands are so special in this context of possessive rule. We are a predatory species, however, and we need to admit that drive to possess without connection—a topological failure. At times this apprehension seems impossible to contemplate, if we wish to make good global decisions—and meanwhile the Lord of the Flies is content to wait.

Where tribes and peoples of the Earth have lived long and with deeply rooted customs, their place had been their Treasure Island, no matter how rich or poor. Nations and countries are symbolic, and their distinct isolation from each other is a matter of peace or war, or something in between. This we see when things go well, but more disturbingly when we observe the fracturing of larger, often artificially created states.

One hopes that cohering fragments may become reasonable flows, but violent realignments are as complicated as particle physics, and there seems no way to counter angry tribal warfare, when it mixes the global hunger for resources and military power and then collaborates with "the pale fiend, cold-hearted commerce." Knowing these sources of instability, we can only return to a rethought artifice of culture, the mapping of a *global topology.* A commercially abstracted electronic map, whose high-speed unified grid is in some ways a surreal fiction, is not an answer to Baskerville or Donne. Their vision we could manage, but only if we trust that connecting paths occur *within* bodies and also *between* bodies, whose larger connection requires a balancing of both types.

There has to be an *entente* between shape and scale, therefore. The primacy of edges in topology tells us that for every edge the *within* and the *between* together imply this balancing act, which is why shape and scale together permit the quality and the quantity of matter to commingle, if there is to be life. In this life-process an ethics of scale adjusts to an aesthetics of shape and lessens the distracting grip of informational glut. The separate pieces, like pieces in a poem by Anne Carson, would create a new *unlost,* and the feel of the fragment, the island as I call this exemplary remnant, will replace the torn page of an old handwritten letter with a reborn leaf. When we are serious about boundary disputes, we inquire into the island dichotomies, relating life and death to a larger union.

All borders are to some extent blurred edges, foldable, and John Donne felt all seven and seventeen types of ambiguity and he knew that if no man and equally no woman or child is an island, we must transform the image and imagery of this limited place we inhabit, to understand better how to use and to accept our planet's borders, boundaries, and divisions—in a word, our limits to every kind of family. Cynics will say we cannot afford to entertain such "unreal" global dreams, we cannot afford to deny our obsession that more is not enough, we cannot afford to balance the treasures of land and sea. Although in metaphor no man or woman is an island and all might in theory have a fair share in the continent, we struggle with too much uncertainty about value and values; the task appears too wide and deep, for as a character asks in Shakespeare's late drama of the impassioned, confused Trojan war, *Troilus and Cressida,* "What's aught, but as 'tis valued?" There are dangerous distances to travel, when linking belief and knowledge and imagination, for their values are a matter of *quality,* not quantity.

Rejecting foolish optimism and weary depression, let us say that a figure such as John Donne is a startling example of the creating, dividing, judging mind, introducing and then balancing doubt. At heart a poet, with capacious rhetoric he could allow a large place for science, although in his time not alone in his willingness to confront our divided and distinguished worlds. Tested by his faith, asking what we can ever really know, Donne was much obsessed with death and the shortness of life. His eloquence shows how much he valued wonderment, when he praised its relevance to understanding human existence and when for the listener he rediscovered the ancient riddle: "Do you know who and especially *where* you are?" This, to conclude, is the explorer's question, and as we have seen, it is not solely a matter of measuring dimensions and sizes—it is a shaping of our consciousness, our changing thoughts, our vanishing dreams, our memories vague or sharp, our aspirations, our fears and regrets and hopes and indeed all our living experience. Let us then suppose that our excesses, though not the balanced usage attending a technological control of material advance, though we are never sure where to place the knife-edge of this perilous balance, would seem to rest on a denial of our recurrent human nature, the recurrence King Lear discovered. Some imbalance may too easily happen with kingship and its "reasons," a breakdown following the labyrinthine path along which John Donne's *Devotions* aimed to navigate safely. He, of course, was imagining a spiritual or religious solute to the riddle. But what if place and position are actually terms for human population identifying its desires solely as situs? What if *analysis situs* is only searching for positions of power—the kind a chess player designs in his chess problems of gaining power to control the end game?

That would require our topology to study islands in the tribes of humanity, as sets of persons defining their countable numbers as qualities of social space. Technology, however simple or sophisticated, too often favors a collapse of reasoned and seasonable value, and with too much tribal isolation Donne could rightly use the phrase, "all coherence gone." A tribal structure, when it exaggerates insular social boundaries and rigid conformity of language, will tend to oppose a free *bricolage,* to use a term invented by Claude Lévi-Strauss. He identified uncritical consistency in society as too much ritual repetition, when such insular positioning adopts an allegorical language, which controls group behavior through unexamined icons. It is almost as if space would then become tyrannical emptiness and in that sense a *puritan* space, as if the group could forever protect its insular identity. Then Donne's hope for a merging continent would be an illusion, and the dream of fixed position would be just that, mere wish.

This book began with a comment on the global anxiety of which our "conquests of nature" appear to be the chief material cause of ominous suspicion. Perhaps, even so, with Donne's hope we may be ensnared by another kind of power, the power of eloquence and rhetoric, blinding us to the actual difficulty of an ecumenical generosity and vigorous global self-restraint. Sadly, perhaps all men are actual mortal islands, as he ruefully wondered. We humans can only ask for whom the bell tolls, since every person's life and the whole biosphere, like Donne's own life and his recovered health, are the climate itself—an emergent occasion, a prayer to the horizon.

When we ask in principle what is this horizon, what meanings it includes, our answer can only amount to an expansive conjecture. Widening our perspective, we can say that beyond the physical sciences horizon calls for studying humanist and humane qualities and values, even though our journey may be obscured by ambiguous metaphors, whose figures of thought are the marks of civilization and general culture.

Topology has shown us the face of *permanence in change*, and horizon is a complex exemplar of this change, since it implies opening and closing the door at the same time. A moving edge enables our sphere to have no edges in an absolute sense, which calls us to live at once on both the inside and outside of a real world, on the subjective inner and objective outer fields of consciousness. While horizon retreats as we advance, it can also turn back towards us, like the tide. If we are born to travel, we must lean upon the sphere and its horizon; we need not quest final knowledge, where we know all things for sure. Thought is indeed "divided and distinguished" and sermons end, so let us imagine the emergent occasion, even should it make for difficult actions and uncertain decisions.

Background Reading

•

Acknowledgments

•

Index

•

Background Reading

These are among the books and articles, several translated into English, which have been useful to me and to which I allude throughout. At times I have eliminated footnotes from my text. Titles with [b] following them have particularly useful bibliographies. Editions listed are often convenient reprints of earlier editions.

Abrams, M. H. *The Correspondent Breeze: Essays on English Romanticism.* New York, 1984.

———. *Natural Supernaturalism.* London, 1971.

Adams, Colin C. *The Knot Book.* New York, 1994.

Agassi, Joseph. *Faraday as a Natural Philosopher.* Chicago, 1971.

Al-Azm, Sadik J. *Kant's Theory of Time.* New York, 1967.

Albert, David Z. *Time and Chance.* Cambridge, MA, 2000.

Alexander, Victoria N. *The Biologist's Mistress: Rethinking Self-Organization in Art, Literature, and Nature.* Litchfield Park, AZ, 2011.

Allemann, Beda. "Metaphor and Antimetaphor." In *Interpretation: The Poetry of Meaning,* edited by Stanley R. Hopper and David L. Miller. New York, 1967.

Andersen, Nathan. "Dynamic Boundaries: Place in Aristotle's Biology." *Graduate Faculty Philosophy Journal* 25, no. 1 (2004): 5–31.

Aristotle. Nichomachian Ethics. Tr. with commentary, H. G. Apostle. Grinnell, IA, 1984.

———. *The Poetics.* Tr. with commentary, Stephen Halliwell. Chapel Hill, NC, 1987.

Auerbach, Erich. *Scenes from the Drama of European Literature.* Minneapolis, 1984.

Bachelard, Gaston. *The Poetics of Space.* Tr. Maria Jolas. Orion Press, New York, 1964.

Badiou, Alain. *Deleuze: The Clamor of Being.* Tr. Louise Burchill. Minneapolis, 2000.

Bailes, Kendall E. *Science and Russian Culture in an Age of Revolutions: V. J. Vernadsky and His Scientific School, 1863–1945.* Bloomington, IN, 1990.

Baskerville, Gordon L. "Advocacy, Science, Policy, and Life in the Real World." *Ecology and Society* 1, no. 1 (1997). http://www.ecologyandsociety.org/vol1/iss1/art9/.

Bassler, O. Bradley. *Diagnosing Contemporary Philosophy with the Matrix Movies*. London, 2016.

———. *The Long Shadow of the Parafinite; Three Scenes from the Prehistory of a Concept*. Boston, 2015.

———. *The Pace of Modernity: Reading with Blumenberg*. Melbourne, 2012.

Bate, Jonathan. *Romantic Ecology: Wordsworth and the Environmental Tradition*. London, 1991.

Bateson, Gregory. *A Sacred Unity: Further Steps to an Ecology of Mind*. Ed. Rodney E. Donaldson. New York, 1991.

Beiser, Frederick C. *German Idealism: The Struggle against Subjectivism 1781–1801*. Cambridge, MA, 2002.

Benjamin, Andrew E., Geoffrey N. Cantor, and John R. R. Christie, eds. *The Figural and the Literal: Problems of Language in the History of Science and Philosophy, 1630 to 1800*. Manchester, UK, 1987.

Berger, Harry, Jr. "Fictions of the Pose." *Representations* 46, Spring (1994): 87–120.

———. *Figures of a Changing World: Metaphor and the Emergence of Modern Culture*. New York, 2015.

Berlin, Isaiah. *The Power of Ideas*. Princeton, NJ, 2002.

———. *The Roots of Romanticism*. Princeton, NJ, 2001.

Bertalanffy, Ludwig von. *Problems of Life: An Evaluation of Modern Biological and Scientific Thought*. New York, 1952.

Berthoff, Ann E., ed. *Richards on Rhetoric: I. A. Richards: Selected Essays (1929–1974)*. New York, 1991.

Blay, Michel. *Reasoning with the Infinite: From the Closed World to the Mathematical Universe*. Chicago, 1998.

Bloom, Harold, ed. *Romanticism and Consciousness: Essays in Criticism*. New York, 1970.

Blum, Jerome. *The End of the Old Order in Rural Europe*. Princeton, NJ, 1978.

Blythe, Ronald. *Akenfield: Portrait of an English Village*. New York, 1969.

Bohm, David. *Wholeness and the Implicate Order*. London, 1980.

Bohm, David, and B. J. Hiley. *The Undivided Universe: An Ontological Interpretation of Quantum Theory*. London, 1995.

Bondi, Hermann. *Relativity and Common Sense: A New Approach to Einstein*. New York, 1964.

———. *The Universe at Large*. Garden City, NY, 1960.

Bonner, J. T. *Size and Cycle: An Essay on the Structure of Biology*. Princeton, NJ, 1972.

Bourdieu, Pierre. *Language and Symbolic Power*. Cambridge, MA, 1991.

———. *Outline of a Theory of Practice*. Tr. Richard Nice. Cambridge, 1979.

Bridgman, P. W. *The Nature of Physical Theory*. Princeton, NJ, 1936.

Brooks, C. E. P. *Climate through the Ages*. 2nd rev. ed. New York, 1970.

Buell, Lawrence. *Writing for an Endangered World*. [b] (See especially Ch. 2, "The Place of Place.") Cambridge, MA, 2001.

Burke, Kenneth. *A Grammar of Motives*. Berkeley, CA, 1969.

Burroughs, John. *Signs and Seasons.* New York, 2008.
Byrd, William. *The Prose Works of William Byrd of Westover: Narratives of a Colonial Virginian.* Ed. Louis B. Wright. Cambridge, MA, 1966.
Byrne, Ruth M. J. *The Rational Imagination: How People Create Alternative Reality.* Cambridge, MA, 2005.
Campbell, SueEllen, et al. *The Face of the Earth: Natural Landscapes, Science, and Culture.* Berkeley, 2011.
Carson, Rachel L. *The Edge of the Sea.* Boston, 1998.
———. *The Sea around Us.* New York, 1951.
———. *Silent Spring.* Boston, 1962.
———. *Under the Sea Wind.* New York, 1991.
Cassirer, Ernst. *The Individual and the Cosmos in Renaissance Philosophy.* New York, 1963.
———. *The Philosophy of the Enlightenment.* Princeton, NJ, 1979.
Casti, John L. *Complexification: Explaining a Paradoxical World through the Science of Surprise.* New York, 1994.
Casti, John L., and A. Karloqvist, eds. *Art and Complexity.* Amsterdam, 2003.
Changeux, Jean-Pierre, and Alain Connes. *Conversations on Mind, Matter, and Mathematics.* Princeton, NJ, 1995.
Chatelet, Gilles. *Figuring Space: Philosophy, Mathematics, and Physics.* Tr. Robert Shore. Dordrecht, NL, 2000.
Chen, Luonan. "Topological Structure in Visual Perception." *Science* 218, no. 12 (1982): 699–700.
Cloudsley-Thompson, John. *Ecology.* London, 1998.
Cohen, Morris R. *Reason and Nature: An Essay on the Meaning of Scientific Method.* New York, 1978.
Coleridge, S. T. *Biographia Literaria.* London, 1975.
Colie, Rosalie. *Paradoxia Epidemica: The Renaissance Tradition of Paradox.* Princeton, NJ, 1966.
Collingwood, R. G. *The Idea of History.* Revised ed. Oxford and New York, 2005.
Colodny, Robert G., ed. *Mind and Cosmos: Essays in Contemporary Science and Philosophy.* Pittsburgh, 1967.
Coupe, Lawrence, ed. *The Green Studies Reader: From Romanticism to Ecocriticism.* London, 2000.
Cowan, George, David Pines, and David Meltzer, eds. *Complexity: Metaphors, Models, and Reality.* Reading, MA, 1994.
Crosby, Alfred W. *Ecological Imperialism: The Biological Expansion of Europe, 900–1900.* Cambridge, 1993.
———. *The Measure of Reality: Quantification and Western Society, 1250–1600.* Cambridge, 1997.
Dantzig, Tobias. *Number: The Language of Science.* New York, 2007.
Davies, P. C. W. *The Forces of Nature.* 2nd rev. ed. Cambridge, 1986.

Davies, Paul. *About Time: Einstein's Unfinished Revolution.* New York, 1996.
Davis, K. S., and J. A. Day. *Water: The Mirror of Science.* Garden City, NY, 1961.
Deacon, Terence W. *The Symbolic Species: The Co-evolution of Language and the Brain.* New York, 1997.
Debord, Guy. *The Society of the Spectacle.* New York, 1995.
Deemter, Kees van. *Not Exactly: In Praise of Vagueness.* Oxford, 2010.
Deleuze, Gilles. *Difference and Repetition.* Tr. Paul Patton. New York, 1994.
———. *The Fold: Leibniz and the Baroque.* Minneapolis, 1993.
———. *Pure Immanence: Essays on Life.* New York, 2005.
Denman, D. R. *Origins of Ownership: Brief History of Land Ownership and Tenure from Earliest Times to the Modern Era.* London, 1959.
Detienne, Marcel. *The Creation of Mythology.* Tr. Margaret Cook. Chicago, 1986.
Deutsch, David. *The Beginning of Infinity: Explanations that Transform the World.* New York, 2012.
Douglas, Mary. *Implicit Meanings: Essays in Anthropology.* London, 2003.
———. *Purity and Danger: An Analysis of the Concepts of Pollution and Taboo.* London, 1966.
Duhem, Pierre. *The Aim and Structure of Physical Theory.* Princeton, NJ, 1982.
Eco, Umberto. *The Role of the Reader.* Bloomington, 1979.
Eddington, A. E. *The Philosophy of Physical Science.* Ann Arbor, MI, 1958.
Edelman, Gerald M., and Giulio Tononi. *Universe of Consciousness: How Matter Becomes Imagination.* New York, 2000.
Edwards, Paul N. *A Vast Machine: Computer Models, Climate Data, and the Politics of Global Warming.* [b] Cambridge, MA, 2010.
Elder, John. *Imagining the Earth: Poetry and the Vision of the Earth.* Urbana, IL, 1985.
Empson, William. *Seven Types of Ambiguity.* First edition 1930. Reprint, New York, 1947.
———. *The Structure of Complex Words.* First edition 1951. Reprint, Ann Arbor, MI, 1967.
Engell, James. *The Creative Imagination: Enlightenment to Romanticism.* Cambridge, MA, 1981.
Evans, John G. *The Environment of Early Man in the British Isles.* London, 1973.
Fabiano, Paolo. *The Philosophy of Imagination in Vico and Malebranche.* Tr. and ed. Giorgio Pinton. Firenze, 2009.
Felstiner, John. *Can Poetry Save the Earth? A Field Guide to Nature Poems.* New Haven, CT, 2009.
Feynman, Richard. *The Character of Physical Law.* Cambridge, MA, 1967.
Field, Thalia. *Point and Field.* New York, 2000.
Fletcher, Angus. "Complexity and the Spenserian Myth of Mutability." *Literary Imagination* 6, no. 1 (2004): 1–22.
———. *Time, Space, and Motion in the Age of Shakespeare.* Cambridge, MA, 2007.

Foster, R. G., and Leon Kreitzman. *Rhythms of Life: The Biological Clocks that Control the Daily Lives of Every Living Thing.* [b] New Haven, CT, 2004.
———. *Seasons of Life: The Biological Rhythms That Enable Living Things to Thrive and Survive.* [b] New Haven, CT, 2009.
Frank, Philipp. *Modern Science and Its Philosophy.* Cambridge, MA, 1949.
Freud, Sigmund. *The Uncanny.* Penguin ed. New York, 2003.
Frye, Northrop. *Myth and Metaphor: Selected Essays.* Ed. Robert Denham. Charlottesville, VA, 1990.
Galison, Peter. *Einstein's Clocks, Poincaré's Maps: Empires of Time.* New York, 2003.
Gamow, George. *One Two Three. . . . Infinity: Facts and Speculations of Science.* New York, 1988.
Gatti, Hilary. *Giordano Bruno and Renaissance Science.* Ithaca, NY, 1999.
Geertz, Clifford. *Local Knowledge: Further Essays in Interpretive Anthropology.* New York, 1983.
Gell-Mann, Murray. *The Quark and the Jaguar: Adventures in the Simple and the Complex.* New York, 1994.
Gibson, James J. *The Ecological Approach to Visual Perception.* Boston, 1979.
Gleiser, Marcelo. *The Dancing Universe: From Creation Myths to the Big Bang.* Hanover, NH, 1997.
Glotfelty, Cheryll, and Harold Fromm, eds. *The Ecocriticism Reader: Landmarks in Literary Ecology.* Athens, GA, 1996.
Goldstein, E. Bruce. "The Ecology of J. J. Gibson's Perception," *Leonardo* 14, no. 3 (1981): 191–195.
Gribbin, John. *Alone in the Universe: Why our Planet is Unique.* Hoboken, NJ, 2011.
———. *In the Beginning: The Birth of the Living Universe.* Boston, 1993.
Gribbin, John, and Mary Gribben. *James Lovelock: In Search of Gaia.* Princeton, NJ, 2009.
Guillen, Michael. *Bridges to Infinity: The Human Side of Mathematics.* Los Angeles, 1983.
Gunderson, L. H., C. S. Holling, and S. Light, eds. *Barriers and Bridges to the Renewal of Ecosystems and Institutions.* New York, 1995.
Hacking, Ian. *The Emergence of Probability: A Philosophical Study of Early Ideas about Probability, Induction, and Statistical Inference.* Cambridge, 1978.
Hadot, Pierre. *The Veil of Isis: An Essay on the History of the Idea of Nature.* Cambridge, MA, 2006.
Halpern, Daniel, ed. *On Nature: Nature, Landscape, and Natural History.* San Francisco, 1987.
Hanson, N. R. *Patterns of Discovery: An Inquiry into the Conceptual Foundations of Science.* Cambridge, 1958.
Harrison, Robert Pogue. *Forests: The Shadow of Civilization.* Chicago, 1993.
———. *Gardens: An Essay on the Human Condition.* Chicago, 2008.

Heisenberg, Werner. *Encounters with Einstein and Other Essays on People, Places, and Particles.* Princeton, NJ, 1983.
———. *Physics and Philosophy: The Revolution in Modern Science.* New York, 1962.
Henderson, Andrea. "Math for Math's Sake: Non-Euclidean Geometry, Aestheticism, and Flatland." *PMLA* 124, no. 2 (2009): 455–471.
Henson, Robert. *The Rough Guide to Climate Change.* London, 2008.
Hiltner, Ken. *Milton and Ecology.* Cambridge, 2003.
———. *What Else Is Pastoral? Renaissance Literature and the Environment.* Ithaca, NY, 2011.
Holland, John R. *Emergence: From Chaos to Order.* New York, 1999.
———. *Hidden Order: How Adaptation Builds Complexity.* New York, 1996.
———. *Signals and Boundaries: Building Blocks for Complex Adaptive Systems.* Cambridge, MA, 2012.
Holton, Gerald. *Einstein, History, and Other Passions.* Reading, MA, 1996.
———. *The Scientific Imagination.* Cambridge, MA, 1998.
———. *Thematic Origins of Scientific Thought: Kepler to Einstein.* Cambridge, MA, 1980.
Huggett, Nick, ed. *Space from Zeno to Einstein: Classic Readings with a Contemporary Commentary.* Cambridge, MA, 1999.
Husserl, Edmund. *Phenomenology and the Crisis of Philosophy.* New York, 1965.
———. *The Phenomenology of Internal Time-Consciousness.* Bloomington, IN, 1966.
Hutchinson, G. E. *The Art of Ecology: Writings of G. Evelyn Hutchinson.* Ed. David. K. Kelly, David M. Post, and Melinda D. Smith. New Haven, CT, 2010.
"Intellect and Imagination." Special issue, *Daedalus: Journal of the American Academy of Arts and Sciences* 109, no. 2 (Spring 1980).
Jackson, J. R. de J. *Method and Imagination in Coleridge's Criticism.* London, 1969.
Jasper, David, ed. *The Interpretation of Belief: Coleridge, Schleiermacher, and Romanticism.* New York, 1986.
Kant, Immanuel. *The Critique of Judgment.* Tr. J. H. Bernard. Amherst, NY, 2000.
———. *Critique of Judgement.* Tr. James C. Meredith. Rev. and ed., Nicholas Walker. Oxford, 2008.
Kellert, Stephen H. *In the Wake of Chaos.* Chicago, 1993.
Kirk, John T. O. *Light and Photosynthesis in Aquatic Ecosystems.* Cambridge, 1944.
Klein, Etienne. *Chronos: How Time Shapes Our Universe.* New York, 2005.
———. *Conversations with the Sphinx: Paradoxes in Physics.* London, 1996.
Klein, Etienne, and Marc Lachièze-Rey. *The Quest for Unity: The Adventure of Physics.* New York, 1999.
Kline, Morris. *Mathematics and the Physical World.* Dover ed. New York, 1981.
Koyré, Alexandre. *Etudes d'histoire de la pensée scientifique.* Paris, 1973.
Kuhn, Thomas S. *The Structure of Scientific Revolutions.* 2nd ed., enlarged. Chicago, 1970.
Landes, D. S. *Revolution in Time: Clocks and the Making of the Modern World.* Cambridge, MA, 2000.

Latour, Bruno. *Politics of Nature: How to Bring the Sciences into Democracy.* Cambridge, MA, 2004.

Lauer, Quentin. *Phenomenology: Its Genesis and Prospect.* New York, 1958.

Leask, Nigel. *The Politics of Imagination in Coleridge's Critical Thought.* New York, 1988.

Lefebvre, Henri. *The Production of Space.* Tr. Donald Nicholson-Smith. Oxford, 1991.

Lem, Stanislaw. *Imaginary Magnitudes.* San Diego, 1984.

Lethaby, W. R. *Architecture, Nature, and Magic.* London, 1956.

Levin, Simon. *Fragile Dominion: Complexity and the Commons.* Cambridge, MA, 1999.

Lewontin, Richard. *The Triple Helix: Gene, Organism, and Environment.* Cambridge, MA, 2000.

Lloyd, G. E. R. *Polarity and Analogy: Two Types of Argumentation in Early Greek Thought.* Cambridge, 1966.

Lorenz, Edward N. *The Essence of Chaos.* Seattle, 1993.

Lovejoy, A. O., and George Boas. *Primitivism and Related Ideas in Antiquity.* Baltimore, 1997.

Lovelock, James. *Gaia: A New Look at Life on Earth.* Oxford, 1987.

———. *Homage to Gaia: The Life of an Independent Scientist.* Oxford, 2003.

———. *The Revenge of Gaia: Earth's Climate Crisis and the Fate of Humanity.* New York, 2006.

Lucas, J. R. *A Treatise on Time and Space.* London, 1973.

Lyotard, Jean-François. *The Inhuman: Reflections on Time.* Tr. Geoffrey Bennington and Rachel Bowlby. Stanford, CA, 1991.

Manuel, F. E., and F. P. Manuel. *Utopian Thought in the Western World.* Cambridge, MA, 1979.

Manuel, Frank E. *Shapes of Philosophical History.* Stanford, CA, 1965.

Margulis, Lynn, and Eduardo Punset, eds. *Mind, Life, and the Universe: Conversations with Great Scientists of Our Time.* White River Junction, VT, 2007.

Margulis, Lynn, and Dorion Sagan. *Dazzle Gradually: Reflections on the Nature of Nature.* White River Junction, VT, 2007.

Matthiessen, Peter. *Wildlife in America.* First edition 1959. Reprint, New York, 1987.

Mazur, Barry. *Imagining Numbers: (particularly the square root of minus fifteen).* New York, 2003.

McKibben, Bill. *Eaarth: Making a Life on a Tough New Planet.* New York, 2011.

———. *The End of Nature.* New York, 1999.

———. *Enough: Staying Human in an Engineered Age.* New York, 2003.

Mellor, D. H. *Real Time.* Cambridge, 1981.

Mendelson, Bert. *Introduction to Topology.* 3rd ed. New York, 1990.

Merchant, Carolyn. *The Death of Nature: Women, Ecology, and the Scientific Revolution.* New York, 1980.

Merleau-Ponty, Maurice. *The Merleau-Ponty Aesthetics Reader: Philosophy and Painting.* Evanston, IL, 1993.

———. *Nature: Course Notes from the Collège de France.* Evanston, IL, 2003.
———. *Phenomenology of Perception.* London, 2012.
———. *The Prose of the World.* Evanston, IL, 1973.
———. *Sense and Non-Sense.* Evanston, IL, 1964.
———. *The Visible and the Invisible.* Evanston, IL, 1968.

Meyer, Steven, and Elizabeth Wilson, eds. "Whitehead Now." Special issue, *Configurations: A Journal of Literature, Science, and Technology* 13, no. 1 (Winter 2005).

Meyrowitz, Joshua. *No Sense of Place: The Impact of Electronic Media on Social Behavior.* New York, 1985.

Miller, J. Hillis. *Topographies.* Stanford, CA, 1995.

Mitchell, Melanie. *Complexity: A Guided Tour.* Oxford, 2009.

Mitchell, W. J. T., ed. *Landscape and Power.* Chicago, 1994.

Moore, Walter. *A Life of Erwin Schrödinger.* Cambridge, 1994.

Morris, Richard. *Time's Arrows: Scientific Attitudes toward Time.* New York, 1985.

Morrison, Philip, Phylis Morrison, and the Office of Charles and Ray Eames. *Powers of Ten: About the Relative Size of Things in the Universe.* New York, 1982.

Morton, Timothy. *Ecology without Nature: Rethinking Environmental Aesthetics.* Cambridge, MA, 2007.

———. *Hyperobjects: Philosophy and Ecology after the End of the World.* Minneapolis, 2013.

———. *The Poetics of Spice: Romantic Consumerism and the Exotic.* Cambridge, 2006.

Mumford, Lewis. *Art and Technics.* New York, 1952.

Myers, Norman. *Ultimate Security: The Environmental Basis of Political Stability.* New York, 1993.

Newman, James R. *The World of Mathematics.* 4 volumes. New York, 1956.

Newton, Roger G. *Galileo's Pendulum: From the Rhythm of Time to the Making of Matter.* Cambridge, MA, 2000.

Nickles, Thomas, ed. *Scientific Discovery, Logic, and Rationality.* Dordrecht, 1980.

Nicolson, Marjorie Hope. *Mountain Gloom and Mountain Glory: The Development of the Aesthetics of the Infinite.* Ithaca, NY, 1959.

Nietzsche, Friedrich. *Friedrich Nietzsche on Rhetoric and Language: With the Full Text of His Lectures on Rhetoric Published for the First Time.* Ed. Sander L. Gilman et al. New York, 1989.

———. *Philosophy and Truth: Selections from Nietzsche's Notebooks of the Early 1870s.* Tr. and ed. Daniel Breazeale. Atlantic Highlands, NJ, 1979.

Oelschlaeger, Max. *The Idea of Wilderness: From Prehistory to the Age of Ecology.* New Haven, CT, 1991.

Ogilvie, Brian W. *The Science of Describing: Natural History in Renaissance Europe.* Chicago, 2006.

Ortony, Andrew, ed. *Metaphor and Thought.* Cambridge, 1982.

O'Shea, Donal. *The Poincaré Conjecture: In Search of the Shape of the Universe.* [b] New York, 2007.

Pagels, Heinz R. *Perfect Symmetry: The Search for the Beginning of Time.* New York, 1986.

Peat, F. David. *Infinite Potential: The Life and Times of David Bohm.* New York, 1997.

Pérez-Gómez, Alberto. *Built upon Love: Architectural Longing after Ethics and Aesthetics.* Cambridge, MA, 2006.

Perspecta: The Yale Architectural Journal. Volume 19. Ed. Brian Healy. Jay Fellows et al., contributors. Cambridge, MA, 1982.

Peterson, David. L., and V. Thomas Parker, eds. *Ecological Scale: Theory and Applications.* [b] New York, 1998.

Pickover, Clifford A. *The Möbius Strip.* New York, 2007.

Poidevin, Robin le. *Travels in Four Dimensions: The Enigmas of Space and Time.* Oxford, 2003.

Poidevin, Robin Le, and Murray Macbeath, eds. *The Philosophy of Time.* Oxford, 1993.

Poincaré, Henri. *The Value of Science: Essential Writings.* New York, 2001.

Polanyi, Michael. *Personal Knowledge: Towards a Post-Critical Philosophy.* New York, 1964.

Popper, Karl. *Objective Knowledge: An Evolutionary Approach.* Oxford, 1972.

Price, Huw. *Time's Arrow and Archimedes' Point: New Directions for the Physics of Time.* New York, 1996.

Putnam, Hilary. *The Collapse of the Fact/Value Dichotomy.* Cambridge, MA, 2004.

———. *Philosophy in an Age of Science.* Cambridge, MA, 2012.

Rajchman, John, and Cornel West, eds. *Post-Analytic Philosophy.* New York, 1985.

Reichenbach, Hans. *The Rise of Scientific Philosophy.* Berkeley, 1951.

Richards, I. A. *Coleridge on Imagination.* Bloomington, IL, 1965a.

———. *The Philosophy of Rhetoric.* New York, 1965b.

———. *Poetries and Sciences* [originally *Science and Poetry,* 1926]. New York, 1970.

Richards, Robert J. *The Romantic Conception of Life: Science and Philosophy in the Age of Goethe.* [b] Chicago, 2002.

Richeson, David S. *Euler's Gem: The Polyhedral Formula and the Birth of Topology.* [b] Princeton, NJ, 2008.

Ricoeur, Paul. *The Rule of Metaphor: Multi-Disciplinary Studies of the Creation if Meaning in Language.* Toronto, 1977.

———. *Time and Narrative.* 2 volumes. Chicago, 1984.

Risi, Vincenzo. *Geometry and Monadology: Leibniz's* Analysis Situs *and Philosophy.* Berlin, 2007.

Rödl, Sebastian. *Categories of the Temporal: An Inquiry into the Forms of the Finite Intellect.* Cambridge, MA, 2012.

Rogers, John. *The Matter of Revolution: Science, Poetry, and Politics in the Age of Milton.* Ithaca, NY, 1996.

Rorty, Amélie Oksenberg, ed. *Essays on Aristotle's Ethics.* Berkeley, CA, 1980.

———, ed. *Essays on Aristotle's Poetics.* Princeton, NJ, 1992.

Rorty, Richard, ed. *The Linguistic Turn.* Chicago, 1971.

Rosenberg, Harold. *Discovering the Present: Three Decades in Art, Culture, and Politics.* Chicago, 1976.

Rousset, Jean, ed. *Anthologie de la Poésie Baroque Française.* Paris, 1968.

Ruelle, David. *Chance and Chaos.* Princeton, NJ, 1993.

Russell, Bertrand. *Essays in Analysis.* Ed. Douglas Lackey. London, 1973.

Sagan, Dorion. *Cosmic Apprentice: Dispatches from the Edges of Science.* Minneapolis, 2013.

Sainsbury, R. M. *Paradoxes.* [See on "Vagueness and the Paradox of the Heap."] Cambridge, 1988.

Sambursky, Samuel. *Physics of the Stoics.* IV. "The Whole and Its Parts." Princeton, NJ, 1987.

Sartre, Jean-Paul. *The Imaginary.* London, 2004.

Scarry, Elaine. *Resisting Representation.* New York, 1994.

Schelling, F. W. J. *The Philosophy of Art.* Minneapolis, 1989.

Schlegel, Friedrich. *Philosophical Fragments.* Tr. Peter Firchow. Fwd. Rodolphe Gashé. Minneapolis, 1991.

Schneider, Eric D., and Dorion Sagan. *Into the Cool: Energy Flow, Thermodynamics, and Life.* [b] Chicago, 2005.

Schneider, Stephen H., et al., eds. *Scientists Debate Gaia: The Next Century.* Cambridge, MA, 2004.

Schrödinger, Erwin. *My View of the World.* Woodbridge, CT, 1983.

———. *Nature and the Greeks and Science and Humanism.* Cambridge, 1996.

———. *Space-Time Structure.* Cambridge, 1994.

———. *What Is Life?* (Includes *Mind and Matter* and *Autobiographical Sketches.*) Cambridge, 1992.

Schuyt, Michael, and Joost Elffers. *Anamorphoses: Games of Perception and Illusion in Art.* New York, 1976.

Serres, Michel. *Hermes: Literature, Science, Philosophy.* Baltimore, 1983.

Shapere, Dudley. *Reason and the Search for Scientific Knowledge.* Dordrecht, 1984.

Simon, Lawrence H. "Vico and the Problem of other Cultures." *Philosophical Forum: A Quarterly* 25, no. 1 (Fall 1993): 33–54.

Skelton, R. A. *Maps: A Historical Survey of Their Study and Collecting.* Illustrated ed. Chicago, 1975.

Sloterdijk, Peter. *Bubbles: Volume I, Microspherology.* Los Angeles, 2011.

———. *Globes: Volume II, Macrospherology.* Los Angeles, 2014.

Smil, Vaclav. *The Earth's Biosphere: Evolution, Dynamics, and Change.* Cambridge, MA, 2003.

Snyder, Gary. *The Practice of the Wild.* San Francisco, 1990.

Society and Nature 1, no. 2 (1992). *The Philosophy of Ecology.*

Soja, Edward W. *Postmodern Geographies: The Reassertion of Space in Critical Social Theory.* London, 1990.

Sorabji, Richard. *Matter, Space, and Motion: Theories in Antiquity and Their Sequel.* Ithaca, NY, 1988.

Sorenson, Roy. *A Brief History of the Paradox: Philosophy and the Labyrinths of the Mind.* Oxford, 2003.

Sossinsky, Alexei. *Knots: Mathematics with a Twist.* Cambridge, MA, 2002.

Stafford, Barbara M. *Visual Analogy: Consciousness as the Art of Connecting.* Cambridge, MA, 1999.

Stewart, Ian. *In Pursuit of the Unknown: 17 Equations that Changed the World.* New York, 2013.

———. *The Mathematics of Life.* New York, 2012.

Szpiro, George G. *Poincaré's Prize: The Hundred-Year Quest to Solve one of Math's Greatest Puzzles.* New York, 2008.

Thomas, Keith. *Man and Natural World: Changing Attitudes in England, 1500–1800.* New York, 1984.

Trudeau, Richard J. *Introduction to Graph Theory.* New York, 1994.

Turner, Victor. *Drama, Fields, and Metaphors: Symbolic Action in Human Society.* Ithaca, NY, 1975.

———. *The Ritual Process: Structure and Anti-Structure.* Chicago, 1969.

Vernadsky, Vladimir I. *The Biosphere.* New York, 1997.

Vico, Giambattista. *The Autobiography of Giambattista Vico.* Tr. Max H. Fisch and T. G. Bergin. Ithaca, NY, 1944.

———. *New Science.* New York, 2013.

———. *The New Science.* Revised translation, 3rd ed., by Thomas Bergin and Max Fisch. Ithaca, NY, 1968.

———. *On the Most Ancient Wisdom of the Italians: Unearthed from the Origins of Their Language.* Tr. and ed. L. M. Palmer. Ithaca, NY, 1988.

———. *La Scienza Nuova.* Ed. Fausto Nicolini. 2 volumes. Biblioteca Filosophia Laterza. Roma, 1974.

Waldrop, Mitchell. *Complexity: The Emerging Science at the Edge of Order and Chaos.* New York, 1993.

Warnock, Mary. *Imagination.* Berkeley, CA, 1976.

Wayne, Don E. *Penshurst: The Semiotics of Place and the Politics of History.* Madison, WI, 1984.

Weeks, Jeffrey R. *The Shape of Space: How to Visualize Surfaces and Three-Dimensional Manifolds.* New York, 1985.

Westling, Louise. *The Logos of the Living World: Merleau-Ponty, Animals, and Language.* New York, 2013.

Weyl, Hermann. *Philosophy of Mathematics and Natural Science.* New York, 1963.

———. *Symmetry.* Princeton, NJ, 1952.

White, Eric C. *Kaironomia: On the Will to Invent.* Ithaca, NY, 1987.

White, Gilbert. *The Natural History of Selborne.* New York, 1899.

White, Hayden. *Metahistory: The Historical Imagination in Nineteenth-Century Europe.* Baltimore, 1973.

———. *Tropics of Discourse: Essays in Cultural Criticism.* Baltimore, 1985.

Whitehead, Alfred North. *Modes of Thought.* New York, 1968.
———. *Process and Reality.* (The Gifford Lectures, 1927-28.) New York, 1979.
Whiteman, Michael. *Philosophy of Space and Time and the Inner Constitution of Nature: A Phenomenological Study.* London, 1967.
Whyte, Lancelot L., ed. *Aspects of Form: A Symposium on Form in Nature and Art.* New York, 1951.
Wilson, E. O. *The Diversity of Life.* Cambridge, MA, 1992.
———. *The Future of Life.* New York, 2002.
Wilson, Eric. *Coleridge's Melancholia: An Anatomy of Limbo.* Gainesville, FL, 2004.
Wise, M. Norton, ed. *Growing Explanations: Historical Perspectives on Recent Science.* Durham, NC, 2004.
Worster, Donald. *Under Western Skies: Nature and History in the American West.* New York, 1992.
Xin Wei Sha. *Poesis and Enchantment in Topological Matter.* Cambridge, MA, 2013.
Yale French Studies. Vol. 49 (special issue), "Science, Language, and the Perspective Mind." Ed. Timothy J. Reis. New Haven, CT, 1973.
Yourgrau, Palle. *Gödel Meets Einstein: Time Travel in the Gödel Universe.* Chicago, 1999.
Zamora, Lois P., and Wend. B. Faris, eds. *Magical Realism: Theory, History, Community.* [b] Durham, NC, 1995.

Acknowledgments

Working essentially in solitude except for my books, I am most grateful to three colleagues who critically supported this project in its final stages: William Flesch, Kenneth Gross, and Gordon Teskey, and throughout to my old friend and fellow worrier, Mitchell Meltzer, whose arguments he rightly may not quite recognize, despite lighting my way, often with jokes that keep me laughing for days. Dorion Sagan persuaded me not to give up on an earlier draft; O. Bradley Bassler guided me through basic topology, while from the outset I was inspired by Donal O'Shea, the distinguished topologist and college president. His encouraging words have been those of a generous critic. To John Irwin, editor of *The Hopkins Review*, I continue to be indebted, not least for publishing earlier forays, and similarly to the learned literary theorist, Marie-Rose Logan, for publishing my essay, "Imagining Earth," in her journal, *Annals of Scholarship*.

Robert Harrison, author of essential books on ways we humans have shaped the environment, remains an inspiring model, while my old friend Francile Downs of Springs, Long Island, kept me alive to the critical state of our endangered wetlands. On a more literary plane Lindsay Waters of the Harvard University Press has been an unfailing ally, and I wish also to acknowledge support from the John Simon Guggenheim Foundation and from the National Endowment for the Humanities, without whose help researchers would be unable to advance the aims of humane learning.

Most of all, I thank my dear wife, Michelle Scissom-Fletcher, for her intuition and equally for her sorely tested patience, as we have moved from one home to another. Michelle waited, and as I worked through many questions, she understood the ambiguities of the task, and she gave me time.

Index